Small Farm Handbook

SECOND EDITION

TECHNICAL EDITORS

LAURA TOURTE

BEN FABER

PUBLICATION 3526

2011

University *of* California
Agriculture and Natural Resources

To order or obtain ANR publications and other products, visit the ANR Communication Services online catalog at http://anrcatalog.ucdavis.edu or phone 1(800)994-8849. You can also place orders by mail or FAX, or request a printed catalog of our products from

University of California
Agriculture and Natural Resources
Communication Services
1301 S. 46th Street
Building 478 - MC 3580
Richmond, CA 94804

Telephone 1(800)994-8849
(510)665-2195
E-mail: danrcs@ucdavis.edu

Publication 3526
ISBN-13: 978-1-60107-698-4
Library of Congress Control Number: 2011921589

This publication has been anonymously peer reviewed for technical accuracy by University of California scientists and other qualified professionals. This review process was managed by ANR Associate Editor for Agricultural Economics Leslie Butler.

4m-pub-2/11-WJC/RW

EDITORS AND AUTHORS

TECHNICAL EDITORS

Laura Tourte, University of California Cooperative Extension Farm Advisor, Santa Cruz, Monterey, and San Benito Counties

Ben Faber, University of California Cooperative Extension Farm Advisor, Ventura and Santa Barbara Counties

CHAPTER AUTHORS

Aziz Baameur, University of California Cooperative Extension Farm Advisor, Santa Clara, Santa Cruz, and San Benito Counties

Steven C. Blank, University of California Cooperative Extension Specialist, UC Davis

Mark Bolda, University of California Cooperative Extension Farm Advisor, Santa Cruz, Monterey, and San Benito Counties

Francine Bradley, University of California Cooperative Extension Specialist, UC Davis

Marita Cantwell, University of California Cooperative Extension Specialist, UC Davis

Carol Collar, University of California Cooperative Extension Farm Advisor, Kings County

Josh Davy, University of California Cooperative Extension Farm Advisor, Tehama, Colusa, and Glenn Counties

Brenda Dawson, Communications Coordinator, University of California Small Farm Program, UC Davis

Maria de la Fuente, University of California Cooperative Extension Farm Advisor, San Benito and Santa Clara Counties

Ben Faber, University of California Cooperative Extension Farm Advisor, Ventura and Santa Barbara Counties

Mark Gaskell, University of California Cooperative Extension Farm Advisor, Santa Barbara and San Luis Obispo Counties

Holly George, University of California Cooperative Extension Farm Advisor, Plumas County

Deborah Giraud, University of California Cooperative Extension Farm Advisor, Humboldt County

Shermain D. Hardesty, University of California Small Farm Program Director and Cooperative Extension Specialist, UC Davis

John Harper, University of California Cooperative Extension Farm Advisor, Mendocino and Lake Counties

Roger Ingram, University of California Cooperative Extension Farm Advisor, Placer and Nevada Counties

Desmond A. Jolly, University of California Cooperative Extension Specialist (retired), UC Davis

Karen Klonsky, University of California Cooperative Extension Specialist, UC Davis

Jim Leap, University of California Center for Agroecology and Sustainable Food Systems Field Production Manager (retired), Santa Cruz

Deanne Meyer, University of California Cooperative Extension Specialist, UC Davis

Elizabeth Mitcham, University of California Cooperative Extension Specialist, UC Davis

Richard Molinar, University of California Cooperative Extension Farm Advisor, Fresno County

Eric Mussen, University of California Cooperative Extension Specialist, UC Davis

James Oltjen, University of California Cooperative Extension Specialist, UC Davis

Barbara Reed, University of California Cooperative Extension Farm Advisor (retired)

Howard Rosenberg, University of California Cooperative Extension Specialist, UC Berkeley

Trevor Suslow, University of California Cooperative Extension Specialist, UC Davis

Etaferahu Takele, University of California Cooperative Extension Area Farm Advisor, Riverside County

Steve Tjosvold, University of California Cooperative Extension Farm Advisor, Santa Cruz and Monterey Counties

Laura Tourte, University of California Cooperative Extension Farm Advisor, Santa Cruz, Monterey, and San Benito Counties

Michael Yang, University of California Cooperative Extension Hmong Agricultural Assistant, Fresno County

BOOK PRODUCTION

Jim Coats, *Principal Editor,* University of California Agriculture and Natural Resources, Communication Services and Information Technology

Robin Walton, *Senior Artist,* University of California Agriculture and Natural Resources, Communication Services and Information Technology

PHOTOGRAPHY

Front cover: Kathy Keatley Garvey; **insets:** Robin Walton (left and center), Mark Gaskell (right).

Back cover: Robin Walton; **insets:** Kathy Keatley Garvey (left and right), Brenda Dawson (center). *Note:* Permitting domestic animals to forage or graze within fields and orchards at various times during the production cycle is a traditional practice, but one that involves many unknowns regarding food safety risks for uncooked produce that is consumed by humans. Growers are advised to conduct an assessment of the potential for contamination of the product and of all food-contact surfaces on a site- and crop-specific basis for each farm operation.

Grower profiles: Brenda Dawson.

PREFACE AND ACKNOWLEDGMENTS

PREFACE

Small-scale agriculture contributes greatly to the vitality and diversity of California's rich agriculture and the state's history. We prepared this second edition of *The Small Farm Handbook* with the intent to provide small farm operators and other readers with a comprehensive, up-to-date resource that will help them better understand the major considerations of small-scale farming. The current volume follows in the footsteps of the 1994 first edition, which, for many years, was the only publication of its kind. Each chapter of the new edition incorporates the distinct voice and knowledge set of its author or group of authors. You can read individual chapters as separate treatments of 11 different subjects or read them together for a more extensive portrayal of the many topics, issues, and facets of California's small-scale agriculture.

ACKNOWLEDGMENTS

This publication is the result of the collaborative efforts of many University of California authors and small-scale farmers, all of whom provided expertise, support, and insight into the complex aspects of California's diverse agriculture. We extend our sincere thanks to all.

Special thanks are extended to the farmers and their families who helped inspire and shape this publication, in particular Emily Ayala, Manuel Bautista, Mike Lee, Dan Macon, Silvia and Frank Prevedelli, and Jay Ruskey.

ABBREVIATIONS AND ACRONYMS

When government agencies and educational institutions are referenced in a publication, the lengthy organizational names practically guarantee an abundance of abbreviations and acronyms. Here are some of the abbreviations and acronyms most commonly used in this book:

AMS Agricultural Marketing Service (USDA)

ANR University of California Division of Agriculture and Natural Resources

ATTRA National Sustainable Agriculture Information Service

AU Animal Unit

AUM Animal Unit Month

CAFF Community Alliance with Family Farmers

CAL EPA California Environmental Protection Agency

CDFA California Department of Food and Agriculture

CFM California Federation of Certified Farmers Markets

COFA California Organic Foods Act of 1980

CSA Community-supported agriculture

Farmer Mac Federal Agricultural Mortgage Corporation

FCS Farm Credit System

FDA United States Food and Drug Administration

FLC Farm labor contractor

HR Human resources

IPM Integrated pest management

IPP Illness prevention program

NAFDMA North American Farmers' Direct Marketing Association

NASS National Agricultural Statistical Service (USDA)

NIFA National Institute of Food and Agriculture

NRCS Natural Resources Conservation Service (USDA)

ORP Oxidation-reduction potential

OSHA Occupational Safety and Health Acts

PACA Perishable Agricultural Commodities Act

RCD Resource Conservation District

RPC Reusable plastic containers

SARE Sustainable Agriculture Research and Education Program (USDA)

SAREP Sustainable Agriculture Research and Education Program (UC)

UC University of California

US EPA United States Environmental Protection Agency

USDA United States Department of Agriculture

VAPG Value-Added Producer Grant (USDA Rural Development)

VRIC Vegetable Research Information Center (UC)

CONTENTS

INTRODUCTION

INTRODUCTION

1
The Vitality and Viability of Small Farms

DESMOND A. JOLLY

It comes as a constant surprise to those unfamiliar with the data that small farms constitute approximately 80 percent of the farms in California, numbering (according to the United States Department of Agriculture [USDA] definition of "small farm") about 60,000 farms. Admittedly, while a good portion of these farms can more appropriately be regarded as rural residential farms, or as the USDA defines them, residential/life-style farms, a substantial portion of them are purposefully engaged in commercial activity with the intent of producing quality products to satisfy the tastes and preferences of potential consumers. Across the United States, one finds a similarly high proportion of small farms—about 80 percent of all farms. Small farms continue to defy predictions of their demise, as predicted by the conventional wisdom of economies of scale.

How have small farms managed to maintain a viable presence in the agricultural landscape? This chapter discusses some of the factors that are associated with a virtual renaissance of small farms over the last half-century. While much of the detailed discussion here will draw on the California experience, parallel developments have occurred in other parts of the United States as California innovations have diffused throughout the country: the California picture is generally descriptive of the recent trajectory of small farms across the United States. We can identify a number of factors and methods that have enabled the continued survival of California's and America's small farm sector.

FACTORS ASSOCIATED WITH SMALL FARM SUSTAINABILITY

In an undifferentiated market with a relatively uniform product, a small farm cannot compete with a large-scale producer, with its army of workers, mechanized production, and commodity marketing system. Small farms have maintained their role in the food and agricultural system by differentiating and diversifying their products according to a variety of attributes, including the relationship between the producer and the buyer, the production practices they employ, and the localness or regionality of the product. As with any operation, dynamism and continued viability derive from a number of key factors—vision, innovation, execution, and opportunity.

In a study to assess California farm operators' perceptions and experiences of success, researchers asked respondents to rate a number of factors in terms of how they contribute to the success of a farm operation. As shown in table 1.1, the farmers rated natural resources highest (8.86 points out of 10), which is a reasonable notion. Generally speaking, an operation that has a rocky soil or some otherwise difficult set of natural resources would face more formidable challenges than an operation that has access to good soils with adequate water in an area with an equable climate. Vision and marketing expertise came next in the ranking, followed by the financial basis of the operation—a low debt load is seen as very important to prospects for success.

Production knowledge and a willingness to try new things also rate highly. The latter can be seen as tied directly to innovativeness. Other interesting factors from this list include location and a network of connections. Observation as well as anecdotal evidence suggests that location plays an important role in small farm viability. But location appears to affect opportunity in concert with other factors, such as vision, willingness to try new things, and networks. That is, in part, why we see the occurrence of dynamic small farm systems in geographic clusters, where access to resources—land, water, and human resources—are reasonably good and where access to markets is also favorable.

Table 1.1. Average ratings of resources needed to create successful farms (n = from 431 to 434)

Resources for success	Mean rating (out of 10 maximum)
Natural resources (soil, water, climate, etc.)	8.86
Marketing expertise	8.32
Vision and strategic thinking*	8.18
Low debt load	8.13
Production expertise	8.09
Willingness to try new things	8.01
Location	7.83
Experience in agriculture*	7.62
Network of connections	7.41
Having access to finance*	7.28

Source: D. A. Jolly and S. Brodt, "Correlates of Small Farm Success," Research Report, UC Davis Department of Agricultural and Resource Economics, August 2005.*

Ratings of these resources significantly correlated with farm size ($p \le 0.05$).

We have touched on some of the factors that, if present in adequate amounts and in combination, favor the success and viability of small farm operations. Some of these pertain to the farm operation—endowment of natural resources and location, for example. With respect to the attributes of the farm operator, we have mentioned vision and a willingness to try new things. But we need to add some additional qualities and attributes that are conducive to a resilient and dynamic operation. Beyond vision and willingness to innovate—to try new things—an ability to plan and manage resources, including time, is almost indispensable. Since resources are by definition more scarce in a small farming system, good managers are keenly aware of the availability of resources over the seasonal calendar of agricultural operations and do not launch a new enterprise that would be assured of failure because of resource constraints that undermine it.

Many agricultural operations are time critical, meaning that risks increase by orders of magnitude once the timing of the operation falls outside the optimal window. This is true of operations such as ground preparation, seeding, weeding, harvesting, and postharvest activities. The lack of sufficient, adequate labor skills and equipment at the critical time can cause losses of both product and money, so a good planner takes risk into account when managing enterprises. Risks emanate from nature, from the product, from the market, and from humans. A good manager draws from experience and from available information, forms reasonably good impressions as to the likelihood that bad events will occur, and either makes contingency plans or changes, perhaps even aborts, plans if the prospects of success look too poor.

A visionary manager also needs to demonstrate patience, because in agriculture most plans require time before they will come to fruition. Thus, in addition to envisioning the eventual outcome of an undertaking, one needs to nourish the intermediate stages and not lose interest in a project in the midstages or during its gestation period. Additionally, success is more probable if someone in the operation pays attention to detail, not overlooking data that emanate from plants, animals, pests, workers, the market, the regulatory or public policy environment—in short, the entire feedback process.

Finally, the state of health and vigor of the farm operator cannot be overemphasized. A sickly, weak farmer is unlikely to have a very successful farm, as farming is often a seven-days-a-week activity with infrequent vacations. Those fortunate enough to be endowed with a strong constitution, all other things being equal, have an advantage.

One can appreciate from the foregoing discussion that the challenges of operating a resilient, dynamic small farm are formidable. That so many achieve this feat is nothing short of remarkable.

MULTIFUNCTIONAL CHALLENGES

Farm operators pay attention to and make critical decisions about myriad factors that affect the long-term sustainability of their operation. Thus, the farm operator assumes a variety of roles, often simultaneously, and makes a multitude of decisions to optimize various outcomes, often in real time. And since what the farmer does today will affect the outcome next season and perhaps for many seasons to come, he or she must assess the likely impact of today's actions on outcomes in future time periods.

The principal roles that the small farm operator fulfills include those of producer, innovator, steward, investor, strategist, and entrepreneur (see Burpee and Wilson 2004, p.15). The operator must be steward for the household's resources as well as those of the farm and ecosystem. As strategist, she or he must assess market opportunities and constraints and must anticipate and react to changes in public policies. The operator must be an entrepreneur, seeking new opportunities, assessing risks and resources, and making decisions about investments. As an entrepreneur, he or she is concerned with making changes intended to increase returns on goods and services, increase productivity, and decrease costs, and he or she must make critical decisions regarding the allocation of scarce resources—land, labor, capital, and the like. She or he must also identify sources of reliable information and pursue cost-effective ways of gaining access to useful information on a host of issues, ranging from new crop varieties to production methods, irrigation, pest management, postharvest techniques, conservation techniques, risk management, and enterprise economics. That so many family-scale farmers successfully juggle the manifold roles and decisions of their farming and ranching operations is a testament to their genius and tenacity.

TRANSFORMING OPPORTUNITIES: SOCIAL CHANGES, POLICY, AND FARM RESPONSE

While changes in the mass marketing and merchandising of agricultural products have seriously challenged the continued viability of small farms, a number of shifts in consumer life-styles and preferences along with changes in public policy have facilitated enhanced opportunities for smaller-scale farms. Changes in the cultural landscape have engendered changes in the agricultural and food marketing landscape. In this section we briefly discuss some of the key social and policy changes in terms of how they have positioned small farms to access new opportunities for revitalization. Those changes include a growing emphasis on and evolving definition of organic production and increased consumer demand for organic products, farmers markets, community-supported agriculture (CSA), niche and minor crops, agritourism, and added-value production.

The resilience and persistence of smaller-scale farms is one of the remarkable features of California, and indeed of U.S. agriculture, during the last half-century. A resurgence of small-scale farming began in the 1960s, roughly half a century ago, with the revolt of young people against standardization and mass culture. It was an offshoot of the general turn toward counterculture life-styles that many saw as affording a greater level of autonomy and control over their lives as well as leading to more wholesome ways of living. Starting in that period, the nation witnessed many movements associated with food and agriculture. There was a movement against food additives, leading to the eventual ban of many substances. There was a movement against pesticides, herbicides, and other chemicals, partly because of their environmental impact and partly because of their impact on human health, including reproduction and genetic inheritance. Not surprisingly, there was a simultaneous movement towards nonchemical (or, more accurately, non-synthetic chemical) food production systems, foremost among which was organic production. Engaged in initially by neophyte back-to-the-land hippies, organic production spread gradually to commercial-scale small farmers and ultimately to mainstream, large-scale agribusiness producers.

Since there were no mainstream distribution channels at first for organic products (mainstream distributors at the time considered organic products to be substandard and unhealthful), a separate distribution system was created. Health food stores and food cooperatives sold "natural" foods and vitamins and, early on, were a mainstay in the marketing of organically produced products. By 1980, advocates for organic foods succeeded in legitimizing organic production with the passage of the California Organic Foods Act of 1980. This legislation established official recognition for organic products and outlined a protocol for their legal production and sale. This in turn created a niche for innovative and enterprising small-scale producers.

The organic agriculture saga illustrates some of the attributes of resilient small farmers—those who can successfully persist in a dynamic and challenging marketplace. Early on, there were no established protocols for production, postharvest management, or marketing. There were few, if any, fact sheets on organic production systems for various crops. There was, however, vision and passion. But inherent in any venture into uncharted territory is risk—a downside risk of losses, economic and otherwise. Indeed, many farmers suffered losses, some significant, in the early years. But those who avoided catastrophic losses were able to benefit from being early adopters. When the market for organic foods exploded in the late 1980s as a consequence of growing consumer concerns about the impacts of chemical materials used in the production of apples on the health and well-being of children, they were well positioned to reap handsome economic rewards and to establish lasting and lucrative marketing relationships with consumers, wholesalers, and retailers. In fact, some small farmers who were early adopters and innovators were able to grow their operations into farm businesses that now gross over a million dollars per year—no longer small farms, in the strict sense of the USDA definition.

Two other key changes in marketing opportunities and production systems facilitated the viability and persistence of California's small farms over the course of the past half-century. The first is the movement toward direct marketing. At its core, direct marketing depends on a relaxation of the strict size and pack regulations of the state code, and that change elicited initial opposition from vested interests. But a coalition of consumer advocates and farmers eventually prevailed, and by the mid-1970s another marketing channel became available to California's small, family-scale farmers. Traditional marketing channels pose significant disadvantages to small-scale producers. Chain stores typically prefer to save on transaction costs by buying in large volumes. They also put a premium on standardization of product and pack. Consistency of supply over the season is another factor they see as a value. (Nowadays of course, with globalization, there is no season—the season for most crops is 12 months of the year.) Produce sold through a packer-shipper typically pays the farmer less than 20 cents of the retail dollar and entails costs for storage, packing, and marketing—and when the market is bad, the farmer often ends up with negative returns. This conventional marketing channel is therefore a high-risk, low-return option for small farmers.

Direct marketing has proven to be a key pillar in the renaissance of California's small farms. The small farm resurgence began with the enabling legislation that permitted direct sales to consumers at farmers markets. For small farmers, this allowed them to avoid the strict cosmetic and size restrictions, promoted cash flow from cash sales to consumers, and restored face-to-face encounters with consumers, permitting direct and immediate feedback. Many consumers perceived the direct-marketed produce as fresher, tastier and more nutritious, and more local. Initially, direct-market prices tended to be lower than for traditional retail. This is no longer always true, but buyers appear to be more motivated by other perceived attributes, depending, of course, on local demographics. Farmers markets allow farmers to experiment with new products and new varieties and to sharpen their entrepreneurial skills. In some places, such as Davis, California, the farmers market has emerged as a centerpiece social institution of the community, encompassing much more than food marketing. The nearly 600 farmers markets in California continue to provide a foundation for the continued vitality of California's small farms. The movement has spread throughout the United States and has grown to more than 6,000 markets nationwide.

Product differentiation and the sale of niche crops, including a proliferation of specialty crops and ethnically oriented products have become traditional approaches for resilient small farmers. California's apple producers, for example, relied on a narrow range of apples up until the mid 1970s—Red and Golden Delicious, Gravenstein, and a few other varieties. When the state of Washington engaged in a massive acreage expansion in the main varieties and undercut the economic viability of California's small-scale apple producers, many predicted the demise of apple production in California. Instead, California's apple producers retooled their orchards, establishing new, less economically threatened varieties such as Granny Smith, Fuji, Mutsu, and countless other varieties, and California's apple production came back from the brink. The mixed greens product, now a mainstay of the salad greens produce category, was an innovation of California's small farms.

Two other marketing innovations worth noting are community-supported agriculture (CSA) programs and agritourism. As in other cases of change, there were innovators and early adopters. Farms such as Full Belly, Good Humus, and Capay Organics in Yolo County were innovators and early adopters of the CSA model. Like other forms of direct marketing, community-supported agriculture establishes a more predictable cash flow and presents a possibility for developing long-term relationships with consumers. Consumers not only come to rely on their weekly or biweekly basket of produce, they also typically develop closer relationships with the farms and farmers, many of them visiting the CSA farm from time to time, particularly on special days promoted by the farm. This link between farm and consumer, between rural and urban, is a vital part of the resilience and resurgence of the small farm. The demand for closer links between urban and rural, between consumers and farms and farmers has fed into another transforming innovation—agricultural and rural tourism, which is another form of direct marketing.

AGRITOURISM

Agritourism operations have been a going concern in California and elsewhere for decades, although they have operated under different rubrics—as U-pick ranches or orchards, dude ranches, or farm bed-and-breakfast operations. But as globalization and mass marketing gained ground in the marketing of agricultural products and with the consequent squeeze on farm gate margins, more and more producers have sought to transform their farms and ranches into destinations for urban consumers, places where visitors can experience agriculture, food production, and a simpler, more basic rural life-style and ambience. The opportunities for visitors have expanded to accommodate a vast range of interests. In fact, these opportunities are limited only by the imagination of the operator and by physical, regulatory, and economic constraints. What is important is that enterprising agricultural producers have realized that the entire assets (viewed in the widest sense) of the farm or ranch can be utilized in enhancing the income-earning and cash flow opportunities of the operation. The role of agritourism varies from operation to operation. In some cases, it is a valued supplement to income derived from production and marketing of commodities. In other cases, agritourism is the major aspect of the operation's income-generating activities.

Agritourism has developed from a curiosity in the early 1990s to a clear option for farmers and ranchers looking for opportunities to enhance economic viability. As in the case of farmers markets and CSAs, many of the early adopters have developed sound agritourism programs, constantly innovating and improving on their product offerings, building up extensive visitor and client bases, and growing their operations over time. Agritourism is increasingly seen as a complementary enterprise to such operations as smaller-scale wine and olive oil production.

The UC Small Farm Program has been greatly involved in supporting and facilitating the growth and development of the agritourism industry in California and the United States through its education and promotion programs. Information and educational materials from the program are available online (http://sfp.ucdavis.edu).

The adoption of agritourism, including nature and ecotourism, is another aspect of direct marketing to consumers, of capturing a greater share of their retail expenditures and helping to maintain the vitality and viability of smaller-scale farming. There are a number of other important alternatives for marketing directly to consumers, including farm stands, tasting rooms, and marketing through the Internet. The important point is the entrepreneurial approach of the farm operator—a willingness to try new approaches in marketing as well as production, and a willingness to bear the risk that comes with trying something new, whether that means adopting a new crop, a new production system, or a new method of marketing the farm or its products.

NONFARM INCOME AND RESOURCES

We need to mention nonfarm resources, both income and human capital, as reservoirs responsible for flows of valuable inputs into small farming. Just as corporations and businesses diversify to avail themselves of upside opportunities and protect themselves against downside risks, small farm households also participate in nonfarm economic and social activities as a way to manage risk and generate income. Often, access to health care, a very expensive service, comes through the outside employment of a member of the farm family. And whereas farm income may fluctuate from season to season and year to year, and cash flow may sometimes lag because of accounts receivable, nonfarm income can be more stable and predictable, enabling the farm household to weather adverse circumstances and tap capital for new or recurrent expenses. Not surprisingly, many small-scale farms that evidence vitality and sustainability have access to nonfarm economic resources. In fact, for many farms nonfarm income is the larger source of income, enabling resilience and innovation on the farm.

The other type of nonfarm inputs into farming and ranching is human capital, not so much in the form of physical labor as in the form of entrepreneurial talent that draws on experiences and knowledge gained in the nonfarm economy. Many of the most noteworthy developments in farm operations over the past 20 or so years have resulted when nonfarmers entered farming and brought with them a wealth of experience in marketing or information technology or other areas common in the urban economy. They have brought fresh perspectives, unhindered by rural tradition and the constraints of cultural inheritance, and as a result have often been innovators in production and marketing. Two books, *Outstanding in Their Fields: California's Women Farmers* and *California's New Green Revolution: Pioneers in Sustainable Agriculture*, profile some of these innovators.

ADDING VALUE

The various forms of direct marketing have provided opportunities for farmers and ranchers to develop added-value products, from cheeses of various types to olive oils, fruit preserves, dried flowers, dried fruit, fibers and clothing from wool and cashmere, and myriad other products. These opportunities would have been much more limited in the mainstream mass market, where national and multinational corporations have the resources to support and project their brands into supermarkets and specialty outlets. Interestingly, many farm-based, added-value products have worked their way into mainstream markets. Trying to hitchhike on the success of this approach, some mainstream corporations are giving their brands names that project a small farm image.

VOLUNTARY ASSOCIATIONS OF FARMERS

Another way family-scale farmers have added to their resilience is by establishing voluntary associations. Through these associations, growers have enabled local and regional branding and marketing efforts. They have also been a good vehicle for communication, among the farmers themselves and between the farmers and external bodies such as local governments, funding agencies such as foundations, and agencies of the USDA. With respect to communication among farmers, one notable example is the decades-old California Small Farm Conference, held annually at different locations in the state (http://www.californiafarmconference.com/), and another is the EcoFarm Conference, held annually at Asilomar Conference Center in Pacific Grove (http://www.eco-farm.org/programs/efc/), both of which facilitate the sharing of success stories, recent applied research, and information related to USDA or Land Grant university program opportunities.

One of the earliest such voluntary organizations in California was Sonoma County Farm Trails, which recently celebrated 25 years of existence. Today there are many local and regional organizations of small growers, including Placer Grown, Calaveras Grown, Capay Valley Grown, Marin Organic, and Napa-Yolano Harvest Trails. Many grower groups have developed maps to help consumers find their farms. These organizations often conduct outreach and fundraising events in and with their surrounding communities. California grower organizations have their counterparts in other states. Clearly these local and regional aggregations of farmers contribute to the viability and resilience of small family farms.

USDA AND THE LAND GRANT SYSTEM

The USDA, which began around the time of the U.S. Civil War, has historically been charged with the responsibility of supporting the modernization of U.S. agriculture and the development of rural America. In addition, the federal government, through the Morrill Act and later the Smith-Lever and Hatch Acts, instituted a network of Land Grant universities in each state that partner with the USDA in its mission, carrying out applied research and extending the fruits of that research to farmers and rural communities. The Land Grant universities were also charged with the development of human capital for agricultural and rural development through resident instruction of undergraduate and graduate students.

The record shows a consistent and remarkable increase in the productivity of U.S. agriculture, with much of that increase deriving from the applied research and extension of Land Grant institutions and the adoption of improved seed, animals, and agricultural practices by growers. The record shows, as well, correlating sharp declines in the overall number of farms and in the population of rural communities. These trends were initially seen as good outcomes: lower-productivity farms were going out of production.

More recently, though, charges of bias against some groups of farmers and in favor of others were brought against USDA, and those charges were legally sustained. In response, Secretary of Agriculture Dan Glickman initiated a series of institutional changes at USDA as a way to change its values and practices. Among the initiatives was the National Commission on Small Farms, empanelled in 1997 to investigate the policies and practices of USDA vis-à-vis the nation's small farms and to recommend changes in USDA's programs, policies, and procedures to enable USDA to better and more equitably serve the large majority of the United States' farms, its small-scale farms. Upon receipt of the Small Farm Commission's report, Secretary Glickman issued a new small farm regulation and directed USDA agencies to re-orient their programs to better serve small farmers.

The result has been a reinvigoration of USDA's approach to applied research and outreach in support of small farmers, and there have been notable efforts by a number of agencies including the Office of Outreach, the USDA Small Farm Program, the Risk Management Agency, the Farm Service Agency, the Agricultural Marketing Service, and the Natural Resources Conservation Service. The Cooperative State Research Extension and Education Service, now called the National Institute of Food and Agriculture (NIFA), coordinates applied research and extension with the USDA's Land Grant university partners, who conduct applied research and extension programs for farmers in the various states. In California, the University of California's Small Farm Program (established in 1979) has benefited from the revitalization of USDA's small farm efforts by partnering with various USDA agencies through grants and partnership agreements that cover a gamut of areas, including food safety, risk management, marketing, agritourism, and outreach to socially disadvantaged farmers and ranchers. And the University of California Cooperative Extension Small Farm Advisors and members of the UC Small Farm Workgroup have exerted strong efforts in applied research and outreach in service of the state's small farmers.

SUMMARY

A resurgence of small farming in the state has occurred over the past half-century, beginning with the social revolution of the 1960s. Experimentation and adoption of new practices, products, and programs, changes in public policy, and changes in consumer life-styles and preferences have all contributed to this resilience and revitalization. A number of risks still face the small farm community—including globalization, increased concentration and monopolization of genetic material and other farm inputs, increasing scarcity of resources, the aging of small farm operators, and public policies focused on regulation of the environment and the workplace—and these will require continued, timely responses. Even so, one can safely predict that there will continue to be a resilient cadre of small-scale farmers in California and the United States who will respond and adapt to new challenges and opportunities, and, in the process, spawn new systems of production and marketing to respond to changing resource constraints and opportunities of the twenty-first century.

RESOURCES AND REFERENCES

Burpee, G., and K. Wilson. 2004. The resilient family farm: Supporting agricultural development and rural economic growth. Practical Action Publishing, Warwickshire, UK. 192 pp.

California Department of Food and Agriculture. http://www.cdfa.ca.gov.

California Federation of Certified Farmers Markets. http://www.cafarmersmarkets.com.

Jolly, D. A., ed. 2005. Outstanding in their fields: California's women farmers. University of California Small Farm Center, Davis. 134 pp. http://www.sfp.ucdavis.edu/docs/publications.asp?view=7.

Jolly, D. A., and S. Brodt. 2005. "Correlates of small farm success." Research Report, UC Davis Department of Resource Economics.

Jolly, D. A., and I. Kenfield. 2008. California's new green revolution: Pioneers in sustainable agriculture. University of California Small Farm Program, Davis. 188 pp.

USDA Agricultural Marketing Service Farmers Markets and Local Food Marketing. http://www.ams.usda.gov/AMSv1.0/FARMERSMARKETS.

USDA National Institute of Food and Agriculture, Family and Small Farms. http://www.nifa.usda.gov/familysmallfarms.cfm.

USDA National Organic Program. http://www.ams.usda.gov/AMSv1.0/nop.

University of California Small Farm Program. http://sfp.ucdavis.edu.

2
California's Small Farms: An Overview

RICHARD MOLINAR, AZIZ BAAMEUR, KAREN KLONSKY, AND MICHAEL YANG

For centuries, small farms and family farms have been an integral part of American agriculture, and their importance continues today. The United States Department of Agriculture (USDA) currently defines a small farm as one that is family operated and grosses less than $250,000 in annual farm gate sales. In California, many family farms operate on relatively small acreages producing high-value row crops such as vegetables and berries, and because of this many of them, while small in acreage, exceed the USDA's figure for gross farm sales. By this standard, California has a lower percentage of farms that are small farms (85%) than the country as a whole (98%), according to the 2007 Agricultural Census (tables 2.1 and 2.2). However, the same census shows that the largest percentage of farms in California and in the six counties listed in table 2.2 fall under the small farm definition, ranging from 78 to 94 percent of all farms in each of those counties.

Small farm characteristics have changed over time. In California, one of the hallmarks of agriculture is its diversity, not just in terms of the types of crops produced and marketed, but also in farm operator demographics, such as ethnicity, gender, age, and style of farm management. This chapter provides a brief overview of some of these characteristics and changes, along with the current status of California's small farms. Based on the USDA's acreage definition of a small farm, the majority of farms in California qualify as small. The percentage of farms in the state deemed small dropped from 100 percent in 1959 to 85 percent in 2007. Comparing 1959 to 2007, the total number of farms decreased by 18 percent and the number of small farms decreased by 31 percent (table 2.1). There was a drop in the overall number of farms between 1959 and 1974, a resurgence until 1987, and then increases in 2002 and again in 2007. The first resurgence may be attributable in part to an influx of immigrants from Laos, Thailand, and Cambodia; the second, to an increase in Hispanic farm operators and hobby or life-style farmers.

MINORITY FARMERS

While the total number of farms in California has declined over the past 20 years, the number of small farms has increased, along with the number of minority-operated farms in all ethnic groups. Most importantly, the number of principal farm operators who are Hispanic increased by 44 percent from 1997 to 2002 and by another 18 percent from 2002 to 2007 (table 2.3). Hispanic farmers represent the largest minority farmer group in all years of the census, reaching a total of 9,118 principal operators in 2007.

The Hispanic farmer group includes those from Mexico and Central America, and their descendants,

Table 2.1. California farms classified as small farms, 1959 to 2007*

Classification	1959	1964	1969	1974	1978	1982	1987	1992	1997	2002	2007
All farms	99,052	80,852	77,875	67,674	73,194	82,463	83,217	77,669	74,126	79,631	81,033
Small farms	99,052	73,697	70,395	56,125	59,625	66,718	66,146	59,852	54,399	67,327	68,536
Small farm % of all farms	NA	91%	90%	83%	81%	81%	79%	77%	73%	85%	85%

Source: USDA National Agricultural Statistical Service (NASS) Census of Agriculture, 1992, 1997, 2002, 2007, and historical summaries.

* The definition of a small farm prior to 1998 was a farm grossing < $100,000. The number changed in 1998 to $250,000.

who came to the United States seeking economic opportunities. The Asian farmer group includes Chinese, Vietnamese, Cambodian, Hmong, Mien, Lao, and others who immigrated to the United States between 1979 and 2005, following the Vietnam War, also seeking economic opportunities. The group also includes Japanese farmers who immigrated to the country earlier in the twentieth century and their descendants. No statistics are available for groups from other countries, such as Armenia, Croatia, Kazakhstan, Russia, and India, but growers in these groups, and their descendants, also contribute to agriculture in California.

The county with the highest number of Hispanic and Asian minority farm operators in 2007 was Fresno, with 12 percent of the state's Hispanic principal farm operators and 24 percent of the state's Asian principal farm operators (table 2.4). Tulare County had the next largest group of Hispanic farm operators and San Diego County had the next largest group of Asian farm operators.

It should be noted that not all small farmers—and in particular not all minority farmers—were included in the USDA 2007 Agricultural Census. This undercount was a combined result of a lower survey response rate for these groups than for the general population of farmers and of difficulties in identifying these groups of farmers. For this reason, researchers generally consider the statistics for minority farmers to be artificially low. Informal surveys of Hmong growers revealed that at least one-third of the growers did not receive 2007 Census forms.

The only data available on minority farmers other than the census data come from a survey of Asian farmers conducted in Fresno County in 1992 and again in 2007. That survey found that 62 percent of Asian vegetable farmers were Hmong, 30 percent were Lao, 3.4 percent were Chinese, 1.5 percent were Mien, and 1.3 percent were Vietnamese (Ilic 1992, Molinar 2007).

Table 2.2. California's small farms: 2007 state total and top counties

Location	All farms	Small farms	Small farm proportion of all farms (%)
California	81,033	68,536	85
San Diego County	6,687	6,290	94
Fresno County	6,081	4,738	78
Tulare County	5,240	4,095	78
Stanislaus County	4,114	3,335	81
San Joaquin County	3,624	2,818	78
Riverside County	3,463	3,165	91

Source: USDA National Agricultural Statistical Service (NASS) 2007 Census of Agriculture.

FARM STRUCTURE

A farm is defined by USDA as a business reporting $1,000 or more in gross annual sales of agricultural products. The National Commission on Small Farms, appointed by USDA in 1997, placed the cutoff between small and large farms at $250,000 in gross annual sales. Therefore, a farm that grosses anywhere between $1,000 and $250,000 is classified as a small farm. In recognition of the broad range of farming operations captured in this definition, the USDA developed a set of farm classifications to disaggregate small farms into several categories: retirement farms, residential farms, and farming occupation (high and low sales).

The small farm categories depend on the operator's major occupation and income relative to poverty levels. The operators of retirement farms are people who have retired from some other occupation, the operators of

Table 2.3. Number of California farms with minority principal operators, 1987 to 2007

	USDA-designated minority group			
Year	Hispanic	Asian	American Indian	African American
2007	9,118	3,684	1,178	320
2002	7,711	3,562	977	278
1997	5,347	3,408	524	277
1992	3,883	3,292	486	253
1987	3,471	3,373	410	309

Source: USDA National Agricultural Statistical Service (NASS) Census of Agriculture, various years.

Table 2.4. California minority principal farm operators by ethnicity and area, 2007

Hispanic		Asian	
Location	Number of operators	Location	Number of operators
California total	9,118	California total	3,684
Fresno County	1,065	Fresno County	866
Tulare County	1,063	San Diego County	255
San Diego County	860	Sutter County	228
Riverside County	500	Tulare County	199
Stanislaus County	438	Riverside County	190
American Indian		**African American**	
Location	Number of operators	Location	Number of operators
California total	1,178	California total	320
San Diego County	119	Los Angeles County	29
Tulare County	111	Riverside County	26
Fresno County	90	Kern County	25
Stanislaus County	54	Fresno County	21
Shasta County	54	San Diego County	18

Source: USDA National Agricultural Statistical Service (NASS) 2007 Census of Agriculture.

residential/life-style farms report having a major current occupation other than farming, and farming occupation farms are run by an operator whose major occupation is farming. Farming occupation farms are further subdivided into those with gross annual sales of less than $100,000 (shown as lower sales in table 2.5) and medium-scale farms with gross annual sales between $100,000 and $249,999 (shown as higher sales in table 2.5). In addition to the small farm classification, USDA further classifies farms with gross sales of more than $250,000 as large family farms if their annual sales are less than $500,000, and very large family farms if their gross annual sales total $500,000 or more. Finally, any farm with a business structure that is a nonfamily corporation, a cooperative, or an operation run by a hired manager is considered to be a nonfamily farm, regardless of its gross annual sales. USDA conducts an annual Agricultural Resources Management Survey (ARMS) that includes information on farm structure. The most recent survey data available, from the 2008 survey, are the source of the data in table 2.5.

Under this expanded characterization, 27 percent of all farms in California are small farms with operators claiming farming as their primary occupation as of 2008. Another 53 percent of California farms are small but are classified as retirement or residential/life-style. In fact, 34 percent of California farms are considered to be residential/life-style, where the primary income of the operator is not from agriculture, implying that most small farms rely on off-farm income (table 2.5). Family farms make up 93 percent of all farms and garner about two-thirds of gross sales. Very large family farms (grossing over $500,000) represent only 8 percent of all California farms but produce over half of the state's agricultural output or value of production. Nonfamily farms represent another 7 percent of farms and produce a substantial one-third of agricultural output. These two categories combined represent 15 percent of all farm businesses in the state and receive 88 percent of gross farm sales (table 2.5).

Looking at the income distribution in 2008, more than three-fourths of California farms grossed under $100,000 from agricultural production (table 2.6). Of the farms grossing less than $100,000, 31 percent showed farming as the primary occupation, 44 percent showed something else as their primary occupation, and

24 percent were retired. In 2008, 84 percent of California farms fit the USDA definition of a small farm, earning less than $250,000 in farm income. Of these, only 6 percent grossed between $100,000 and $250,000, and 93 percent of those farms defined as small farms actually grossed less than $100,000. Of the farms grossing between $100,000 and $250,000, almost two-thirds identified farming as their major occupation. The other third of the farms were divided between retired farmers and operators reporting something other than farming as their primary occupation.

RESOURCES AND REFERENCES

Ilic, Pedro. 1992. The Southeast Asian Farmers in Fresno County status report 1992. University of California Cooperative Extension, Fresno County, Fresno. Unpublished.

Molinar, Richard H., and M. Yang. 2007. Update: The Southeast Asian Farmers in Fresno County status report 1992 and 2007. University of California Cooperative Extension, Fresno County, Fresno. Unpublished.

Table 2.5. Distribution of California farms, value of production, and farm assets by farm type, 2008

	Farms		Value of production			Farm assets		
Farm type	Number	%	$ per farm	All farms $1,000	%	$ per farm	All farms $1 million	%
Small family farms	65,227	80	40,648	2,651,348	8	1,053,652	68,727	47
Retirement	15,145	19	24,609	372,703	1	996,570	15,093	10
Residential/life-style	27,742	34	21,189	587,825	2	966,666	26,817	18
Farming occupation								
Lower sales	19,696	24	55,669	1,096,457	3	1,052,499	20,730	14
Higher sales	2,644	3	224,797	594,363	2	2,301,900	6,086	4
Commercial family farms	10,210	13	1,969,108	20,104,594	58	5,213,705	53,232	36
Large	3,308	4	395,636	1,308,764	4	2,988,057	9,884	7
Very large	6,902	8	2,723,244	18,795,830	54	6,280,417	43,347	30
Nonfamily farms	6,063	7	1,937,169	11,745,056	34	4,102,753	24,875	17
All farms	81,502	100	423,325	34,501,845	100	1,801,663	146,839	100

Source: USDA Economic Research Service (ERS). Farm Business and Household Survey Data, Customized Data Summaries from ARMS, California 2008, Farm Business Balance Sheet and Farm Income Statement by Farm Typology. http://www.ers.usda.gov/Data/ARMS/app/default.aspx?survey=FINANCE

Table 2.6. Distribution of California farms by gross farm income and operators' primary occupation, 2008

Gross farm income	$1,000,000 or more	$500,000 to $999,999	$250,000 to $499,999	$100,000 to $249,999	Less than $100,000	All Farms
Number of farms	5,428	3,412	3,936	4,816	63,910	81,502
Operator occupation (%):						
Farming	92%	87%	80%	62%	42%	50%
Something else	7%	10%	11%	20%	43%	36%
Retired	1%	3%	9%	19%	15%	14%

Source: USDA Economic Research Service (ERS). Farm Business and Household Survey Data, Customized Data Summaries from ARMS, California 2008, Structural Characteristics by Economic Class. http://www.ers.usda.gov/Data/ARMS/app/default.aspx?survey=FINANCE

Small Farm Handbook

BACKGROUND SKILLS AND KNOWLEDGE

3
Requirements for Successful Farming

RICHARD MOLINAR, AZIZ BAAMEUR, AND KAREN KLONSKY

Successful farmers combine their production knowledge and skills with top-notch business plans and marketing approaches to provide the nation and world with its food. They work with natural resources and manage them to produce high-quality, diverse, nutritious, and abundant products. Farming takes hard work, planning, discipline, and prudent management, and then still more hard work. A farmer must be prepared to devote many hours to raising and selling his or her products, and must be able to deal with the fact that income will vary from year to year depending on the weather, pests, and market conditions. Farmers must be astute and able to manage financial, cultural, and other aspects of farming. Risk is inherent in farming; success is in part the result of identifying and maximizing opportunities while minimizing constraints and losses. A climatic or economic factor that causes disaster in one production region may actually benefit another region or even your own farm.

This chapter will not tell you what agricultural products are the most profitable for your farm. Each farmer or rancher, parcel of land, and marketing strategy is different from every other one, and no single prescription is guaranteed to tell you what enterprise or activity is right for you. Any new farming operation or enterprise that you undertake will, however, require an investment of your time, money, and other resources. Risks will always be involved.

Most successful small farmers are innovative and flexible. They obtain the necessary equipment, use appropriate cultural operations, understand business practices, follow the market closely, watch for new trends, and are astute entrepreneurs. They know that crop quality is important and that it comes into play from the moment they begin planning for their next year's activities. They continue to learn, experiment, develop new skills, and adjust practices to new conditions as they arise. For example, cultivation of blueberries in California's Central Valley was almost unheard of 15 years ago, but then a few farmers, in collaboration with University of California Cooperative Extension Advisors, began experimenting with production and marketing of the crop. Today they are very successful. Successful small farmers are also on constant lookout for other potential new crop, marketing, and value-added opportunities.

Beginning farmers may decide to start their operation as a part-time business, maintaining another off-farm full-time job while they get started. This helps them finance the new business and provides income during the start-up phase. If you begin instead as a full-time farmer, you will need sufficient funds to carry your business and living expenses for 1 to 2 years or until you begin to profit from the sale of your crops, livestock, or other products.

COMMITMENT TO FARMING

If you are considering farming, carefully identify and evaluate your vision, goals, and objectives for your farm. Your reasons for farming may be financial, they may be personal, or they may be a combination of the two. You may

- wish to be your own boss or an entrepreneur in full control of your business
- come from a farming background or inherit a farm
- enjoy farming and want to raise your family in an agricultural setting
- hope to increase or diversify your sources of income
- wish to take advantage of investment incentives in the tax laws

Owning and operating a farm business requires stamina, fortitude, and creativity. Farmers need to be able to withstand the vagaries of product successes and failures and to find creative solutions to unforeseen problems that are sure to arise. Most farmers work between 50 and 70 hours per week. However, operating your own farm allows you some control of your schedule and may mean that family members can take part in the business. It is important to identify and set personal and family goals and objectives in advance to help you make sound decisions and avoid confusion and conflict. Decisions that affect the farm will also affect the family.

If you plan to live on your farm, you need to consider such basics as location and availability of services. What are the advantages and disadvantages of a rural versus an urban setting? Among other things, considerations include how to handle repairs to your home and farm buildings, the cost of electricity or propane, availability of telephone and Internet service, availability and quality of water, quality of and distance to medical care, schools, and stores, quality of roads to and on the farm, characteristics of the local community and neighbors, and accessibility to friends. If you live on a farm in a rural area, you will need to be well organized to minimize the number of trips to the nearest town or supply center, since each trip requires both time and money. Your costs for amenities such as food, fuel, and other consumer goods may be higher in rural than in urban or suburban areas with chain stores. You may decide to rent farm land for the time being and then purchase land later, or you may decide to live in one location and commute to the farm. Transportation and communications technologies allow for a number of alternatives.

Before beginning any farming operation, though, there are fundamental questions you should ask yourself—and honestly answer. These questions may seem superfluous or difficult, but the answers may hold some of the keys to your success. Information offered in other chapters of this handbook will help you find your answers.

Personal questions:

- Have I managed a business before?
- Have I formulated and documented goals and objectives with desired outcomes?
- Do I have the skills and experience necessary for this enterprise?
- Do I have enough personal energy?
- Can I count on my family members for support?
- Does the probable work load correspond with the time of year I want to work and the amount of time I want to work?
- Do I have the time required to develop and launch a new farm, crop, or animal enterprise?
- Will a new enterprise complement my current enterprises?
- Do I care what the neighbors think about my farm or new enterprise?

Finacncial questions:

- Have I established business and marketing plans and strategies that include projected financial statements and budget projections for the first 1 to 3 years?
- Do I have sufficient cash and capital assets to begin a new business?
- Do I have access to credit or loans to assist with the new enterprise?
- Do I have experience managing records and finances, or can I afford to hire a skilled person to do that?

Legal and regulatory questions:

- What are the legal and regulatory requirements for farming in general?
- What are the local laws and ordinances with which I must comply?
- What permits or documentation must I have if I want to apply pesticides?
- Do I need licenses to sell my products at certified farmers markets or other direct-market venues?
- Do I want my farm products to be certified organic? If so, what legal requirements must I follow?
- What kind of a business structure is best for my farm: LLC, partnership, or some other arrangement?
- What kind of insurance do I need or am I required to have?

Land and production questions:

- What is the availability and cost of rental land or land for purchase?
- What soil type(s) are found on the prospective farming area, and is the soil suitable for farming and for the crops I want to grow?
- What plants and animals are adapted to this climatic region?
- What water resources are available for irrigating crops or watering livestock?
- How much water is available, and what is the cost per acre-foot, and is it suitable for my purposes?
- What are the historical and seasonal rainfall patterns for the area?
- What is the water drainage like?
- How might a flood or a drought affect my enterprises?
- Do I want other concurrent uses for the land besides farming, such as wildlife conservation, fishing, or hunting?
- What local environmental constraints, restrictions, or permits might I need to consider?
- Are there any legal or regulatory restrictions on land use?

Building and machinery questions:

- Are existing buildings and facilities adequate to manage the proposed farm operation?
- Can additional storage facilities or other buildings be rented or leased?
- What machinery is on hand, and is any of it required for time-sensitive farm operations?
- Have I got sufficient funds to purchase needed machinery?
- Can machinery be rented, leased, or borrowed?

Labor needs questions:

- How much labor will existing or new enterprises require?
- Will I need additional labor beyond what my family can provide?
- Do I like to supervise people or have I got sufficient resources to hire a farm manager?
- What is the source and availability of local labor?
- Is part-time or seasonal labor available?
- How much will labor cost per hour, per month, or per year?
- Is affordable local housing available for agricultural workers?
- Can I house workers on my farm? Are there legal and zoning restrictions that apply?

Marketing questions:

- What are the market trends and consumer demand for the products I wish to provide?
- Who or what are my target markets and market segments?
- How big is the potential customer base?
- Where do my potential customers live? How will location influence my ability to sell to them?
- Will I sell directly to local consumers, for example, through farmers markets or Community Supported Agriculture (CSA)?
- Will I work with distributors or sell to wholesale markets?
- What seasonal price ranges and fluctuations might I expect?
- What quality standards must I meet for the type(s) of marketing I choose?
- What are the postharvest, food safety, and legal considerations associated with marketing?
- What story can I tell about my farm? Do I want to develop a brand for my farm or my products?

KNOWLEDGE AND SKILLS

Farming involves more than planting, cultivating, and harvesting crops. It requires record keeping, financial forecasting, marketing decisions, handling legal and regulatory issues, dealing with farm credit, and technical knowledge about equipment, plant and animal growth, chemical use, and soils. You can acquire knowledge through written materials, the Internet, workshops and courses, consultation with veteran farmers and other agricultural experts, and hands-on experience. Written material is available at libraries and university and college campuses where agricultural courses are taught. Sources of expertise on farming include other farmers, UC Cooperative Extension Farm Advisors and Specialists, professors, agricultural consultants, business representatives, the California Department of Food and Agriculture (CDFA), and the United States Department of Agriculture (USDA). Many community colleges offer short courses in small-scale and beginning farming. Also, various community-based organizations offer some technical assistance. The Internet is a tremendous source for information. It is best, though, to practice caution with respect to the quality of some of the information available on the Internet. Field days, conferences, and workshops are also organized and offered by UC Cooperative Extension and various other organizations, agencies, packing houses, and nonprofit groups.

Experience is a good teacher. It is important to experiment with new crops or products on a small scale for potential, but you can get better, more useful experience working with a seasoned farmer. Other farmers are an excellent source of information and advice. Many are willing to share their experiences and challenges. Some farmers offer apprenticeships, which may include room and board. A few organizations and schools also offer apprenticeships. Conferences, workshops, and field days are frequent and offer excellent opportunities to meet and network with other farmers and agricultural experts. A good source of information about these opportunities is your local UC Cooperative Extension office or Web site.

Skills that are typical of most successful farmers include a natural affinity for growing crops or raising animals, along with mechanical, business, and people skills. Business, marketing, and communication skills are critical to farming in the twenty-first century and should not be overlooked.

LOCATION AND RESOURCES

If you are not already farming, finding the right property and location should be a high priority for you. Is the land you are interested in suitable for farming? Obtain information about climate, temperature, wind, soils, water quality and availability, drainage, land use laws, possible restrictions and regulations, and the surrounding area. Has the land been surveyed? Are corners and boundaries defined? Is fencing necessary, and if so how much? Consider taxes, including special district and water taxes. What condition are buildings in? How much will improvements cost? Is farm labor available? Are roads in good condition? How far away are the potential markets? Can your product be transported easily?

Is there an urban population close to your farm, and what effects might this have on your operation? Neighboring urban populations may be a potential market but may also be problematic because of possible aversions to certain aspects of farming, including noise, dust, odors, and the use of chemicals. Carefully consider both the advantages and the disadvantages.

Since you will need water for crop, livestock, and domestic purposes, you should consider both the quantity and quality of water. Is it obtainable from surface sources or from underground? Some areas have adequate well water. If land around you has been developed for housing or industrial use or if planning for such use is under way, how does that affect your water supply and when? Contact the local irrigation district about deliveries, hookups, future supplies, and projected costs. Will you have rights to use water from streams or to drill a well? Ownership of water rights is complicated. Sometimes well water can be used for irrigation, but you will want to weigh costs carefully to decide whether returns are likely to justify the costs of drilling and energy investments for pumping. Are ponds an option for water storage, irrigation, livestock, wildlife, or even fish production? Are there springs? What time of year do they flow? Is the water safe to use? Are there regulations and restrictions on its use? Check on the spring again in late summer before you firm up plans to use this source of water.

Farm wastes and their disposal have become increasingly important, with a greater focus on environmental quality, laws, and restrictions, and with urban populations coming closer to the farm. Explore how on-farm waste will be handled. What types of waste will the farm generate? Can the wastes be recycled on the farm? Will they be disposed of off-site? What are the environmental implications of producing or disposing of the waste? Did the previous owner have waste disposal problems? Are latent problems likely to emerge? For example, if pesticides or fuels have been stored on the farm, they may have leaked from underground fuel tanks and other containers, creating polluted soil and water that are expensive to clean up. In some counties, fuel tanks no longer in use must be filled or removed. Chapter 4 contains more information on how to evaluate locations.

MARKETING

Before you decide which crops to grow, you need to develop a marketing plan. This should include a determination of whether you will grow and market organic or conventional products, who you expect will buy your products, the volume they are likely to purchase, and the price per unit that you can expect. You can determine all of this through market research. Also consider your transportation needs or how you might deliver products. What are the distances to market and what will transportation cost? Do you plan to sell your products through distributors and wholesalers, terminal markets, cooperatives, farmers markets, on the farm, or directly to restaurants? The answers to these questions will help you decide what to produce, when to produce it, and the volume of products that the market will bear. Many small farmers prefer to direct-market their products at farmers markets and roadside stands and directly to restaurants. They can get as much as two or three times the money for each pound of fresh market berries sold at a roadside stand rather than to a commercial processor.

Direct marketing is also of great importance to small farmers with highly specialized and perishable products or when volumes are too small for you to sell through conventional channels. According to one farmer, "The easiest part is to make a product that tastes good. The hardest part is distribution and sales. You may have to create demand for your product. You have to go out and sell the product before you produce it. Get your markets, get your shelf space." An excellent source of information is the USDA Market News Service, which gives prices paid at various terminal markets, helping farmers determine what their crop might be worth at the local level. Both current and archived reports are available at the following Web site: http://www.ams.usda.gov/AMSv1.0/marketnews. Chapter 7 will also provide you with marketing information and a greater understanding of markets and marketing.

KEYS TO SUCCESS

What, how, and when to grow and market. A small farmer's economic success depends on his or her knowing what, how, and when to grow and market a quality crop or animal. Careful fiscal management, strong personnel management skills, professional business relationships with markets and consumers, and a willingness to be innovative are also essential elements of success. California farmers produce an extraordinary array of products that are high both in quality and in quantity. Building, selling, and retaining market share are therefore critical to success.

Diversification of products and markets. To minimize risk and operate viable farm businesses, small farmers must diversify their products and marketing approaches in one or more ways. This could include, for example, the production of numerous crops or animals, direct marketing, use of an alternative production or marketing system (i.e., organic or specialty crop production), adding value to products, or formation of a buying or marketing cooperative.

Diversification may allow you to capture a higher percentage of the total amount consumers are willing to pay by direct marketing a variety of complementary products, have year-round income, and establish a niche in the market. Diversification may also include creating an additional on-farm business—for example, value-added jams, agritourism, or a bed-and-breakfast operation—to help augment farm income.

Community partner. Farming on a small scale can be enjoyable and profitable. Local and family farms are often seen as important community partners, providing employment, agricultural products, and other services to the public. You will have a greater chance for success if your farm's presence and its operations are acceptable to your neighbors and the community. Farm odors (such as manure), early morning and late night tractor noise, pesticide applications, and clouds of dust can be objectionable to people who do not understand farming operations and needs. Your success in the community will be improved if you work cooperatively with your neighbors, keeping your self informed about issues and topics of interest to the community and making informed decisions in managing your business.

Hard work. Farmers often work from dawn to dusk in the field, marketing products, and managing the business. Ten- to twelve-hour days are not uncommon. A Central Valley farmer selling at San Francisco Bay Area farmers markets, for example, may spend all day

Friday harvesting and packing 10 to 20 different crops, then arise at 1:00 a.m. Saturday to prepare for market, travel 3 hours to one farmers market, and work half or all of the day there. Afterwards he or she will find an inexpensive motel for the night and then travel the next morning to a nearby Sunday market, driving back to the farm when that market is through.

UNIVERSITY OF CALIFORNIA COOPERATIVE EXTENSION

The University of California Cooperative Extension performs applied research, education, and outreach in local communities throughout the state. UC Cooperative Extension Farm Advisors work to enhance California's agricultural productivity and competitiveness. Together with farmers, industry consultants, government agencies, and nonprofit organizations, they identify current and emerging agricultural trends, opportunities, and problems. Farm advisors collaborate with UC Berkeley, Davis, and Riverside campus-based Cooperative Extension Specialists and Agricultural Experiment Station scientists to research, adapt, and field-test agricultural improvements and solutions and to extend research findings to the public. In addition to Farm Advisors, UCCE has Advisors in nutrition, family, and consumer sciences and the 4-H Youth Development Program, based in more than 50 county offices statewide.

UC COOPERATIVE EXTENSION SMALL FARM ADVISORS

Many UC Cooperative Extension Farm Advisors assist small family farmers in their geographic area and their area(s) of expertise. There are five who are designated specifically as Small Farm Advisors, and direct their work towards small-scale farming.

Aziz Baameur, serving Santa Clara, Santa Cruz and San Benito Counties
Telephone: (408) 282-3127; Fax: (408) 298-5160;
E-mail: azbaameur@ucdavis.edu

Mark Gaskell, serving Santa Barbara and San Luis Obispo Counties
Telephone: (805) 788-2374; Fax: (805) 781-5940;
E-mail: mlgaskell@ucdavis.edu

Manuel Jimenez, serving Tulare County
Telephone: (559) 684-3316 Ext. 216;
Fax: (559) 733-6720; E-mail: mjjimenez@ucdavis.edu

Richard Molinar, a UC Cooperative Extension Small Farm Advisor in Fresno County, advises potential new farmers in his area to try to identify several potential market possibilities before they start a new enterprise. "Marketing is one of the key issues that needs to be nailed down as soon as possible and may even be the determining factor in what you grow. The owner of a specialty retail store in the Bay Area may tell you he doesn't need Chinese cabbage but would definitely buy Shanghai bok choy from you. Or by contacting a restaurant chain in Sacramento, you may find out that they would be very interested in Chinese eggplants."

Molinar also emphasizes that the small farmer should be a persistent squeaky wheel. "There are many resources available free of charge to the farmer," he says, "and you should consult as many of these resources as possible. Examples might include pump testing from programs at an agricultural college or university, soil testing or evaluation from USDA Natural Resources Conservation Service (NRCS) or UC Cooperative Extension offices, grant dollars to conduct sustainable agriculture research on your farm from USDA Sustainable Agriculture Research and Education (SARE) program, weed, insect, disease, and nematode identification and testing from UC Cooperative Extension, low-interest loans and disaster assistance from USDA Farm Service Agency, information on farm housing and structures from USDA Rural Development, and additional information that may be helpful from other organizations including CAFF (Community Alliance with Family Farmers), Farm Bureau, California FarmLink, Lao Family, and other community-based organizations. Cost-share programs for on-farm conservation projects may also be available through such organizations as local Resource Conservation Districts." See the list of additional resources at end of this chapter.

Ramiro Lobo, serving San Diego County
Telephone: (858) 694-3666; Fax: (858) 694-2849;
E-mail: relobo@ucdavis.edu

Richard Molinar, serving Fresno County
Telephone: (559) 456-7555; Fax: (559) 456-7575;
E-mail: rhmolinar@ucdavis.edu

You may also contact other UC Cooperative Extension Farm Advisors at any of the county UC Cooperative Extension offices listed in the Appendix.

SELECTED ORGANIZATIONS

African American Farmers of California
3171 W. Kearney Boulevard
Fresno, CA 93706-2214
Telephone: (559) 442-1893
E-mail: AaFarmers97@aol.com

California Association of
Pest Control Advisers (CAPCA)
1143 N. Market Boulevard, Ste. 7
Sacramento, CA 95834
Telephone: (916) 928-0705
http://www.capca.com

California Farm Bureau Federation
Statewide office:
2300 River Plaza Drive
Sacramento, CA 95833
Telephone: (916) 561-5500
http://www.cfbf.org

California FarmLink
PO Box 2224
Sebastopol, CA 95473
Telephone: (707) 829-1691
http://www.californiafarmlink.org

Community Alliance with Family Farmers (CAFF)
36355 Russell Boulevard
PO Box 363
Davis, CA 95617
Telephone: (530) 756-8518
http://www.caff.org

Fresno Center of New Americans
4879 E. Kings Canyon Road
Fresno, CA 93727
Telephone: (559) 255-8395
http://www.fresnocenter.com

Lao Family Community of Fresno, Inc.
4903 E. Kings Canyon Road, Ste. 281
Fresno, CA 93727
Telephone: (559) 453-9775
http://www.laofamilyfresno.org

Lao Family Community of Stockton, Inc.
1808 Country Club Boulevard
Stockton, CA 95204
Telephone: (209) 466-0721
http://www.laofamilyofstockton.org

Merced Lao Family Community, Inc.
855 W. 15th Street
Merced, CA 95340
Telephone: (209) 384-7384
http://www.laofamilymerced.com

National Hmong American Farmers, Inc.
2904 N. Blackstone Avenue, Ste. A2
Fresno, CA 93703
Telephone: (559) 225-5309
http://www.nhaf.org

Pesticide Applicators Professional Association (PAPA)
PO Box 80095
Salinas, CA 93912-0095
Telephone: (831) 422-3536
http://www.papaseminars.com

RESOURCES AND REFERENCES

ATTRA National Sustainable Agriculture Information Service. http://attra.ncat.org.

California Department of Food and Agriculture (CDFA) Marketing Division: http://www.cdfa.ca.gov/mkt Organic Program: http://www.cdfa.ca.gov/is/i_&_c/organic.html.

Colorado State University Extension Boulder County Small Farm Program. http://www.coopext.colostate.edu/boulder/ag/smallfarms.shtml.

Cornell University Small Farms Program. http://www.smallfarms.cornell.edu.

Flint, M. L. 1998. Pests of the garden and small farm: A grower's guide to using less pesticide. Second edition. University of California Division of Agriculture and Natural Resources, Oakland. Publication 3332. http://anrcatalog.ucdavis.edu/SmallFarms/3332.aspx.

George, H., and E. Rilla. 2005. Agritourism and nature tourism in California. University of California Division of Agriculture and Natural Resources, Oakland. Publication 3484. Pp. 159. http://anrcatalog.ucdavis.edu/SmallFarms/3484.aspx.

Lobo, R., D. Wallace, K. Robb, and S. Parker. 2006. San Diego County agricultural directory & guidelines for agricultural enterprises. UC Cooperative Extension San Diego County, San Diego, CA. http://ucce.ucdavis.edu/files/filelibrary/2017/11766.doc.

Lyman, A. 1993. Personnel decisions in the family farm business. University of California Division of Agriculture and Natural Resources, Oakland. Publication 3357. http://anrcatalog.ucdavis.edu/SmallFarms/3357.aspx.

Molinar, R., and M. Yang. 2001. Guide to Asian specialty vegetables in the Central Valley, California. UC Cooperative Extension Fresno County, Fresno, CA. http://cefresno.ucdavis.edu/files/4902.pdf.

———. 2004. Family farms in Fresno, California. UC Cooperative Extension Fresno County, Fresno, CA. Publication OPA21. http://cefresno.ucdavis.edu/files/4900.pdf.

———. 2007. Small farm resource directory. UC Cooperative Extension Fresno County, Fresno, CA. http://cefresno.ucdavis.edu/files/35589.pdf.

Myers, C., et al. 1998. Specialty and minor crops handbook. University of California Division of Agriculture and Natural Resources, Oakland. Publication 3346. http://anrcatalog.ucdavis.edu/SmallFarms/3346.aspx.

Oregon State University Extension Service. Small Farms. http://smallfarms.oregonstate.edu.

USDA National Institute of Food and Agriculture Family and Small Farms. http://www.nifa.usda.gov/familysmallfarms.cfm.

University of California Cost and Return Studies. http://coststudies.ucdavis.edu.

University of California Fruit and Nut Research and Information Center. http://fruitsandnuts.ucdavis.edu.

University of California Integrated Pest Management Program (IPM). http://www.ipm.ucdavis.edu.

University of California Postharvest Technology. http://postharvest.ucdavis.edu.

University of California Sustainable Research and Education Program (SAREP). http://www.sarep.ucdavis.edu.

University of California Vegetable Research and Information Center (VRIC). http://vric.ucdavis.edu.

University of Florida Small Farms and Alternative Enterprises. http://smallfarms.ifas.ufl.edu.

University of Maryland Small Farm Outreach and Technical Assistance Program. http://extension.umd.edu/about/1890/programs/smallFarmer.cfm.

Varea Hammond, S., and L. Tourte. 2004. Agricultural Resources Directory for Monterey, San Benito, and Santa Cruz Counties. University of California Cooperative Extension, Monterey and Santa Cruz Counties.
English: http://cesantacruz.ucdavis.edu/files/25110.pdf.
Spanish: http://cesantacruz.ucdavis.edu/files/20465.pdf.

Washington State University Small Farms Team. http://smallfarms.wsu.edu.

4

The Basics

AZIZ BAAMEUR, JIM LEAP, AND RICHARD MOLINAR

Launching a new enterprise or farm takes imagination, resolve, and execution. Imagination resides in a person's ideas, goals and objectives. Why do you want to farm? What purpose will farming satisfy? What value will the farm provide you, your family, and your community? Resolve is the physical and mental capacity to engage in strategy, planning and development. Are the land, soil, and water resources of appropriate quality and are they suitable or sufficient for farming? What is the history of the land or existing farm? Are there limiting factors or constraints that you will want to consider? What zoning laws or local ordinances affect or impact farming in the area? What are the products you want to grow or raise? What is your potential market and what price might you expect to receive? What capital and cash resources do you have available to begin this enterprise or farm?

Execution is the final step—implementation—when the goals, objectives, planning, and development stages transition into specific steps that lead to the realization of the endeavor. In this chapter, we will explore various aspects of the basics associated with beginning a new enterprise or farm. Part of this exploration will require you to gather and analyze information and facts specific to your own situation.

LAND

Very few people are lucky enough to have access to an existing family farm. In most cases, prospective farmers will need to search for a parcel of land suitable for farming. Take time to consider the following points.

Price. Quality farmland in close proximity to lucrative markets is very expensive to purchase, if it is available at all. Coastal property prices are particularly high relative to, for example, many areas in California's Central Valley. Farmland near urban areas is generally more expensive than land in rural areas. Hilly terrains are almost always less expensive than flat ground, but may also come with special production constraints. Renting or leasing ground may be the only viable option available to you. The California Chapter of the American Society of Farm Managers and Rural Appraisers performs an annual survey of rent and lease values for the state (http://www.calasfmra.com/). You may also contact your local University of California Cooperative Extension Farm Advisor. A good strategy might be to find property that is for lease with an option to buy, since that arrangement could justify any improvement expenses you may incur. New and beginning farmers who wish to lease or buy agricultural land may also want to contact California FarmLink (http://californiafarmlink.org), an organization that seeks to link aspiring farmers with retiring landowners to sustain family farms and conserve farmlands in the state.

Taxes. In addition to the price of land, consider local taxes. Are taxes based on agricultural value or the potential value of development? Local zoning laws can to a great degree determine your property taxes.

Zoning permits and right-to-farm ordinances. Several zoning laws regulate farming activities. Inquire about the long-range general plans of the community where the land is located. Research current zoning regulations not only for the land you are interested in, but also for surrounding properties. You want to be aware of any prospective zoning changes that may interfere with or hinder your future farming business. Zoning regulations evolve continuously.

Farming in agriculturally zoned areas is the best option. This may reduce the permitting and regulatory hurdles all farmers face as part of farming in today's business environment. Ask if your county or area has a right-to-farm ordinance that recognizes and supports farming as an important and valuable part of the community. Right-to-farm ordinances serve to educate community members and can help minimize the objections neighbors may raise with respect to noise, dust, odors, and various perceived risks associated with farming activities. They also afford farmers some protection from nuisance complaints or legal actions. It is a good idea to get an idea of the prevailing outlook of residents and business leaders in the surrounding areas where you plan to farm.

Williamson Act. Zoning that gives special tax consideration to farm land is termed the Williamson Act. The California Land Conservation Act of 1965, also known as the Williamson Act, was enacted with the purpose of preserving agricultural land. The act enables owners of private land to voluntarily enter into contracts with counties and cities to restrict the development of agricultural land in consideration for lower tax levels. Contracts are negotiated in 10- to 20-year increments or until either party breaks the contract by filing a notice of nonrenewal, thereby removing the land from special tax consideration. It is important to note that not all counties offer Williamson Act contracts. The economic climate may also determine whether Williamson Act tax considerations will apply.

To take advantage of the Williamson Act, the farm must meet certain conditions:

- The total land area must be no less than 100 acres, *or*
- If two or more owners are involved, they can combine parcels smaller than 100 acres to meet the 100-acre requirement.

Also, a smaller unit holding may be granted an exemption to the 100-acre requirement if the parcel is deemed by the county to have unique agricultural characteristics. The seat of authority for these decisions varies from county to county, but the county's planning department can be a good source of information and guidance.

The Williamson Act periodically undergoes review and is amended. It is best to check with the Department of Conservation (http://www.conservation.ca.gov) to learn about its current status.

Conservation easements, land trusts, and mitigation. Landowners can voluntarily sell or donate future development or subdivision rights for their land to public agencies or private organizations in the form of conservation easements or land trusts. In this sort of arrangement, the owner generally forfeits personal ownership of those land rights, but in exchange protects the land and its resources for agricultural use for a period of time, often in perpetuity. Mitigation requires that farmland, when converted to a nonagricultural use (i.e., development), be compensated for by protecting equivalent or comparable land elsewhere. Easements, trusts, and mitigation arrangements vary depending on state and local policies and laws. For more information, contact the American Farmland Trust (http://www.farmland.org) and the Land Trust Alliance (http://www.landtrustalliance.org).

Parcel size. Parcel size can, to some extent, determine the scope and the type of an enterprise or farm operation. Capital and cash resources, management skills, and labor requirements are other critical determinants. Marketing strategies need to be closely aligned with production goals for the grower to understand and realize the income potential of an operation. For animal production, the land's carrying capacity (the number of animals it can support per acre) must be considered. Some farmers find that the best approach is to start small, working only a part of their land. They then build on their experience and success to expand their businesses. No one enterprise or parcel size can guarantee profitability to all farmers.

SOIL

Soil type, quality, and health. The type, quality, and health of the farm's soil affect crop production and farming strategies. California is home to a wide variety of soil types. The USDA classifies soils in classes numbered I through VIII. Soils at the lower categories, I to III, are determined to have few or no limitations for producing agricultural crops. These soils tend to have deep profiles, good water infiltration, and minimum runoff. Those in the higher classes, IV to VIII, require that you place more restrictions on farming activities. USDA Natural Resource Conservation Service (NRCS) can provide information, maps, and other resources to help you understand your soil and its potential for farming. Many UC Cooperative Extension offices also offer this information.

An understanding of soil quality is an important part of farming. Soil quality involves such characteristics as texture, structure, depth, water-holding capacity, drainage, salinity, pH, nutrient availability, and tilth.

- *Soil texture* refers to the amount of sand, silt, and clay present in a soil profile. A soil with high sand content will be termed light, while one with more clay is called heavy.
- *Soil structure* refers to the aggregation of soil particles and the soil's ability to withstand erosion.
- *Soil depth* refers to the soil profile and the depth available for plant root growth.

- *Water-holding capacity* of a soil refers to the maximum amount of water it will retain after drainage of excess water due to the force of gravity.
- *Drainage* is the removal of excess water and dissolved salts from the soil profile.
- *Soil salinity* is determined by measuring the soil's content of water-dissolved electrically charged ions, which are characteristic of salts. Some of the most common salts are chlorides, sodium sulfates, carbonates, bicarbonates, magnesium salts, and potassium salts. Salt ions in the soil are the result of many processes such as natural weathering of soil minerals, water and fertilizer applications, and upward migration from groundwater.
- *Soil pH* is an index of a soil's acid or alkaline (basic) nature. The pH range runs from 0 (very acidic) to 14 (very alkaline), with a neutral value of 7. A wide variety of plants thrive in soils with a pH of 6.0 to 8.0. Soil pH values found to be at either extreme (high or low) have a negative influence on nutrient availability for most plants.
- *Nutrient availability* refers to the presence and accessibility of sufficient soil nutrient levels for adequate plant growth.
- *Soil tilth* refers to a soil's structure and aggregates and its ability to be worked without losing its structural stability.

Soils are considered healthy when they have well-balanced or optimal physical, chemical, and biological properties that support sustained crop growth and development over time with few or no limitations. This includes suitable levels of soil organic matter, nutrients, and populations of macroscopic and microscopic organisms. In addition to sustained crop production, soils with balanced physical, biological, and chemical properties also tend to have good water penetration and infiltration characteristics and good water-holding capacity. If you are not familiar with your soil's characteristics, it is a good idea for you to collect samples and submit them to one or more laboratories in your area for analysis. The testing results will give you an important tool for informed crop production decisions. You will find additional soils information in Chapter 9, Growing Crops.

A QUICK SOIL TEST

Soil test results should include these data:

- pH: soil acidity or alkalinity levels
- Salinity: Electrical conductivity (EC) and concentration of the ions chloride (Cl^-), sodium (Na), magnesium (Mg), calcium (Ca), boron (B)
- Cation-exchange capacity (CEC): CEC is the measure of a soil's ability to hold and release positively charged nutrients (cations) at a given pH. It is an indicator of soil fertility commonly expressed in milliequivalents per 100 grams of soil (meq/100 g).
- Fertility: nitrogen (N), phosphorus (P), potassium (K).

WATER

The availability and quality of water are important basic factors you will want to consider. Before buying or renting land, ask the following questions: Is there enough water of good quality to produce crops or maintain livestock? What is the source of water for the farm? Are there wells, and if so, how deep are they? What is the pumping cost per acre-foot? Are there special fees or charges? Are there streams or ponds on the property that I can use? What are the environmental restrictions or regulations that I need to consider? Do not forget to compare the availability, quality, and cost of water from different sources. Access to a primary and secondary source of water is an advantage that should not be overlooked.

For instance, for most crops you will need a water flow of 5 to 15 gallons per minute (gpm) for each acre of land to be cultivated. In addition to water for growing crops, you may need to consider how much water you will need for washing and cooling produce, dust control, and protecting crops from cold and frost.

Water quality. Water quality affects crop growth from planting to harvest. Poor-quality water may cause failure of seeds to germinate, scorching of plant leaves, or, in the worst-case scenario, crop failure. Water quality can be determined by a laboratory's analysis. If the lab results are baffling to you because of new or unfamiliar terminology, make sure to ask the lab to explain anything that is unclear or confusing.

Water quality is judged based on the amount of dissolved solids or salts in the irrigation water. Both surface (i.e., delivered, stream, or lake) and well water contain salts at varying levels. Also, some surface and well water may have been exposed to contaminants such as pesticides, fertilizers, or other wastes. In high amounts, salts can damage and even kill plants and plant roots. The quantity of dissolved solids determines the suitability of the water for crop use. Some crops are more sensitive than others to water containing salts. Please see Chapter 9, Growing Crops, for additional information on water and irrigation.

WATER QUALITY ANALYSIS

The lab report usually lists the overall water salinity assessment in terms of electrical conductivity (EC). Pure water will not conduct electricity, whereas water with dissolved salts will conduct in proportion to its total dissolved salts (TDS) content.

EC is expressed in decisiemens per meter (dS/m) or millimhos/centimeter (mmhos/cm) (both units are equivalent to one another).

TDS is expressed in parts per million (ppm).

Generally speaking, EC values lower than 1 will not be problematic in crop production. There are salt-tolerant crops that can produce well under moderate water salinity conditions.

CLIMATE

In a narrow sense, climate refers to the average weather conditions in an area over a given period of time. It is advisable to consider the climatic assets or limitations of the area that you are considering for your farm. What is the length of the growing season? Does the area have microclimates within larger dominant climatic zones? What is the average cumulative heat unit total (degree-days) per season for each area? What is the monthly temperature profile? What is the average annual rainfall? Are there storm events each year? If so, how intense are the events, and how many are typical per year? What impact might the storms have on soil quality, erosion potential, management, and crop production? Is the area windy? If so, how frequent and severe are the winds? The answers to these questions will help you plan your operation and minimize the risks associated with adverse climatic conditions.

Temperature. Temperature and weather conditions determine not only the rate of plant growth but also the development of crop pests and beneficial organisms. Plants are divided into cold-season and warm-season plants. This distinction is not exact; there is a graduation of sensitivity to cold. Plants are thus categorized as cold hardy, cold semihardy, and cold sensitive. Besides affecting the plants themselves, excessive cold and heat can restrict human agricultural activities. Adequate chilling hours are also important for tree fruit production and fruit quality. It is important to search for information on recorded minimum temperatures. Also important is the number of days per year that the temperature does not exceed 30°F. For information on chill requirements, refer to the UC Davis–based Fruit and Nut Research and Information Center (http://fruitsandnuts.ucdavis.edu/Weather_Services/About_Chilling_Units_&_Hours.htm). Some temperature-related restrictions can be alleviated if you follow certain practices that effectively extend the growing season.

Season extension. Special agricultural practices have been developed that can extend a farm's growing season by adding time either at the beginning or the end of the season or by reducing the length of time the plants actually have to be in the ground. Transplants, plastic mulch, plastic tunnels, raised beds, row or floating covers, greenhouse structures, and heating or frost protection all can serve to extend the season.

Pest management. Climatic data from local weather stations and online weather networks can also be used to forecast pest conditions and plan for pest management.

The goal is to use one or more databases to predict the growth and development of insects (both pests and beneficial) and the development of diseases. Working together, growers and scientists have developed computer models for the development of powdery mildew and botrytis in grapes, black mold and late blight in processing tomatoes, botrytis in strawberries, and scab and fire blight in apples and pears. Please consult Chapter 9. Growing Crops, for additional information.

Climate change and agriculture. The topic of climate change is inescapable. It has been debated at universities, in the popular media, and in public and private forums. Climate change is affected by the interaction of a variety of factors, including atmospheric levels of carbon dioxide (CO_2), temperatures, and rainfall or storm events. It is difficult to predict how climate change might impact agriculture at this time. Current scientific models indicate a number of potential outcomes, including reduced agricultural productivity due to lower precipitation rates and a rise in agricultural pests, diminished soil fertility because of excess rainfall, increased water runoff and soil erosion, and a shift of agricultural production to new regions or locations because of temperature changes. This will be a topic for research and discussion far into the future.

EQUIPMENT

Appropriate equipment is essential to farming. It is also expensive to buy and maintain. Growers may borrow equipment from one another or custom hire people who own specialized or expensive equipment. Equipment rental is another option. Buying used equipment will help keep your initial investment costs low, but this approach has both disadvantages and advantages. Include equipment needs in your farm plan. Make a list of essential items and get them first. Acquire other items as your budget allows or as you find good deals.

Tractor selection for a diverse row crop farm. When you are growing a diverse mix of row crops, there are many factors you need to consider in selecting the right tractor for your farming operation. The most important questions to consider are what configuration, what size, what horsepower (HP), what make, and whether to buy the tractor new or used. To make the best decision in tractor selection, one must first determine the scale of the operation: small-scale, mid-scale, or larger-scale small farm.

Small-scale farm (market garden). Small-scale growers producing mixed vegetables on less than 5 acres are typically limited to a single tractor due to capital constraints and the size of their operation. The drawback to using only one tractor on a small, diverse farm is that the various field operations may require different characteristics in tractors, and it is often difficult to find one tractor that will effectively meet all of the different requirements. Many small-scale operators rely heavily on the use of hand tools such as push planters and wheel hoes, thus eliminating much of the need for in-field tractor operation after the beds have been formed.

Mid-scale farm. For mid-scale operations, in the 5- to 15-acre range, the best option is often to have two tractors: a dedicated tillage tractor for pulling discs, chisels, and various secondary tillage implements, and a second, smaller tractor for operations like planting, cultivating, some secondary tillage, and mowing. At this scale, most implements would be in the 6-foot-wide range.

Larger-scale farm. For farms of 25 acres and more, there are options that might include a relatively new row crop tractor configuration known as a mudder, or high-clearance, tractor. Mudder tractors are commonly used to pull harvest equipment through wetter fields on large commercial farms. They combine good traction and four-wheel-drive capabilities with good crop clearance. They also have fairly narrow tires and wider tire spacing to match typical row crop spacing (up to 80 inches measuring from the center of one tire to the center of the opposite tire). These tractors are most often available in the 60 to 80 HP range and can perform tillage and post-tillage row crop functions fairly efficiently. At this scale, implements can be up to 12 feet wide.

Buying new or used? If you have a good mechanical aptitude, enjoy tinkering, have ready access to tools and information, and have adequate space for working on tractors, then by all means buy used. If these attributes do not describe you, purchasing new equipment might be your best option. Buying used equipment will often save you money in the short term. In addition, it may allow you to find equipment configurations that meet your needs. For example, many of the small-scale specialty row crop equipment configurations are no longer available new in the United States, though they are still being made for European and Asian markets. These small tractors may still be available on the used market for a reasonable price. If you are looking for a vegetable row crop tractor, it is often advisable to look in vegetable growing regions for the makes and models that you like.

Buying new has many advantages as well. Often, new tractors are fairly reasonably priced (especially when compared to buying a new car), and you can often special-order features like creeper gearing, specific tire widths, tire spacing, and auxiliary hydraulics. The other real advantage to buying new is that the newer diesel engines are quieter, cleaner burning, more fuel-efficient, and require less maintenance than their older counterparts. As a general rule, newer tractors are better ergonomically designed than many of the older tractors. This is important, considering that it is not uncommon for an operator to spend a significant amount of time on the tractor.

TIPS FOR BUYING A USED AGRICULTURAL TRACTOR

When buying a used diesel tractor, one of the most important things to remember is that there is probably a reason why the tractor is being sold. There is a good chance that it has reached a point in its life where the amount of wear exceeds the owner's financial ability to maintain it. If the tractor has a functional hour meter with a reading in excess of 10,000 hours, this is usually a good indicator that the tractor is worn out. If you are considering purchasing a used tractor, there are some things you will want to assess carefully. There is expertise involved in assessing a used tractor, so consider hiring an experienced tractor mechanic to inspect it in advance of your purchase. If you are going to do the assessment yourself, you will want to consider the following points:

- Start and drive the tractor and, if possible, attach it to an implement and work it long enough to bring it up to operating temperature. By working the tractor you will be putting it under load and this condition will be very helpful in determining what type of condition the tractor is in.
- Basic indicators of a worn-out tractor include
    - excessive exhaust gases escaping from the crankcase oil filler cap when removed while the engine is operating (this condition is referred to as blow-by)
    - hard starting, black exhaust (especially under load)
    - excessive engine heat under load
    - difficulty shifting
    - excessive gear noise
    - difficulty braking
    - excessive oil leaks from engine and transmission seals
    - low oil pressure
    - weak hydraulics, as indicated by hydraulic lift arms that drop quickly from the raised position when the tractor is shut off

A good way to determine the health of a three-point hydraulic system is to hook up to a heavy implement and see if the tractor is able to pick up the load at idle. Diesel engines can knock somewhat, especially when cold, so it can be a real challenge to discern the sounds of engine bearing failure simply by listening. Diesel engines also turn the crankcase oil black fairly quickly after an oil change, so the appearance of the crankcase oil can also be misleading. A real plus when purchasing a used tractor is if the seller can provide maintenance records. Many farm tractors are poorly maintained, and poor maintenance can significantly limit the useful life of a tractor. The condition of the air cleaner can be a good indicator of how well the tractor has been maintained.

BUILDINGS AND ROADS

When you evaluate a farm for purchase or to rent, consider how much you will have to spend to make buildings and roads fit your operation. Existing buildings, if they fit in your operation and are of high enough quality to be of service, would certainly save on your investment expenses. However, rehabilitating or even disposing of old structures may be more costly than building new ones. Buildings have various uses such as housing animals or equipment, or storage and processing activities. In many instances they are used to provide livestock with shelter from predators.

Road conditions and erodibility may become a greater concern in the wet season. Will area roads be passable in all seasons? Will you need to build roads on your property? The USDA NRCS (http://www.nrcs.usda.gov) can provide help with road and building plans and with conservation. Plans for farm buildings and other structures are also available in some local UC Cooperative Extension offices. UC ANR provides a free 23-page publication, *Rural Roads: A Construction and Maintenance Guide for California Landowners,* that can be downloaded from the UC Division of Agriculture and Natural Resources (ANR) catalog Web site (http://anrcatalog.ucdavis.edu/FarmManagement/8262.aspx). Finally, the Santa Cruz County RCD published information on road construction and management in their newsletter "News from the Road" (various issues) and their 2004 "Private Roads Maintenance Guide for Santa Cruz County" (http://www.rcdsantacruz.org, located in the resources section).

LABOR

Probably the most costly and critical aspect of a farming operation is labor. USDA reports indicate that on the average, 40 to 70 percent of the cost of production goes to labor (Agricultural Labor Management, http://www.cnr.berkeley.edu/ucce50/ag-labor/). The more specialized and unique the operation is, the higher the labor costs.

In planning for an existing operation or a future farm, figuring out your labor needs. How much, when, and where is really only the beginning of what you need to know. Issues of management, contracts, labor regulations and laws, hiring, and firing become all the more critical.

While most agriculture operators are experienced and knowledgeable in the areas of plants, pests, water, and environmental management, few are trained in the area of human resources (HR). It may be surprising to hear HR brought up for a small operation, but employee supervision and management are skills that allow an owner or farm manager to direct and guide a workforce, no matter how small, to be productive. Matching labor needs and the right human resources is the secret to successful management. Concern for the productivity and welfare of employees goes hand in hand. In other words, good management should benefit both the employer and the employee. The greatest efficiency can be obtained from employees who are well trained, motivated, and permitted to take initiative. This is true whether the employer hires workers directly or through a labor contractor.

In addition to the human resources side of labor management, there are legal factors that govern the relationship between an employer and an employee. Myriad laws and regulations related to labor and management are constantly evolving and changing, baffling many would-be experts. The laws and regulations set standards for terms of employment relating to wages, workplace health and safety, break and meal times, and the employment of minors. They also regulate benefits and retirement standards. Labor laws also govern certain employer-employee interactions such as hiring, firing, and the handling of complaints on a variety of topics, including allegations of discrimination. And because of agriculture's reliance on seasonal workers, there are specific regulations to address issues involving authorized workers, temporary agricultural workers, and questions of immigration and nationality. Chapter 8, Labor Management, provides greater detail on many aspects of farm labor.

ENERGY

Farming, like so many other economic activities, is energy driven. With high energy costs and concern about using nonrenewable sources of energy, it has become not only economically sound but of paramount importance that farmers use energy wisely. Energy savings can be realized in every aspect of the farming endeavor.

Buildings. There are many building designs that improve energy usage and save money. From site selection to building materials and design, options are available to help you save energy. Building size and the need for supplemental heating, cooling, or lighting all can be manipulated to minimize energy use and cost. Adequate insulation of buildings can lower energy costs. For example, many different insulation materials are available to help reduce thermal energy loss by conduction, convection, and radiation. Keep in mind such alternative technologies for building materials as thick earthen walls, agriculture fiber such as straw, mineral wools such as rock wool, and fiberglass. Adding passive solar collectors can help you take advantage of the sun's power. In other instances, planting fast-growing trees to provide shade may help reduce cooling costs.

Fuel. In addition to using fuel-efficient gasoline and diesel machinery, farmers are also evaluating alternative biofuels. There are two generations:

- First-generation biofuels rely on sugars, starches, vegetable oils, and animal fats used as fuels for conventional existing engine technologies. Many information resources are available on the Internet.
  - *Vegetable oil,* used mainly with old diesel engines and mostly suitable for warm climates.
  - *Biodiesel fuel* is derived from used animal or vegetable oil and is similar in composition to diesel. The potential for biodiesel fuel is great because around 80 percent of commercial truck and buses run on conventional diesel fuel.
  - *Bioalcohols* are produced by fermenting plant residues. Fermentation is carried out by microorganism and enzyme activity.
  - *Butanol,* also called biogasoline, is a product still in its early stages. It is intended to be used in gasoline engines and will in theory provide a more efficient energy conversion.
  - *Bioethanol,* or *ethanol,* is an alcohol fuel that is the product of enzymatic digestion of starchy and sugary plant materials such as sugar beet, sugar cane, corn, and wheat.
  - *Biogas* (mostly methane) is the gaseous component of the breakdown of organic matter in anaerobic conditions. The parent organic matter can be plants or waste material.
- Second-generation biofuels are based on lignocellulosic biomass feedstock and will require advanced technical methods for production.
  - Technologies currently under research: switchgrass, stover, and wood material broken down to fermentable sugars using specific enzymes.
  - Newer and not-yet-available technologies: biohydrogen (hydrogen production via biological processes), biobutanol, biodimethyl ether, mixed alcohols, and synthetic gas.

Solar power. Current technology uses photovoltaic (PV) cells or arrays to capture solar energy and convert it directly to electricity. The electricity thus produced is used to power equipment or to recharge storage batteries. Solar energy is currently used for irrigation water pumping, livestock watering systems, greenhouse heating, and the processing of agricultural products.

California Senate Bill 1 (SB 1) and the Public Utilities Commission have authorized the California Solar Initiative program, which pays incentives for properly installed and maintained solar power systems. The level of the incentives is based on the level of electrical production, in kilowatts. For additional information consult the U.S. Department of Energy site on frequently asked questions regarding solar energy (http://apps1.eere.energy.gov/solar/cfm/faqs/). Financial incentives may also be available for the installation and use of solar power systems. Find information on solar energy tax credits online at http://www.energy.ca.gov. In addition, commercial alternative energy companies may have lease arrangements available.

Wind power. Wind energy can be a low-cost source of renewable, clean energy. Unlike sunlight, wind is intermittent, even in high-wind corridors. Windmills tend to be noisy, though newer technologies are less so. Critics may point out environmental problems (e.g., incidence of bird killings) or aesthetic problems. The entire United States has been mapped according to wind power capability, and the information is available on the Internet. Also, the U.S. Department of Energy maintains a Web site with frequently asked questions about wind energy (http://www1.eere.energy.gov/windandhydro/faqs.html).

Another Web site for estimating energy requirements associated with irrigation, tillage, nitrogen, and propane costs is available through USDA NRCS (http://www.nrcs.usda.gov/technical/energy). Wisely used, renewable energy can augment conservation efforts and help preserve environmental quality. It is estimated that 90 percent of the energy used in an irrigated farming operation is associated with water use. Energy savings in the processing and pumping of water could immediately save 10 percent of that, and possibly more over time. For more information consult the California Energy Commission (http://www.energy.ca.gov).

CONSERVATION AND ENVIRONMENTAL QUALITY

Conservation and environmental quality are increasingly important aspects of farming. It is important that veteran and prospective farmers be knowledgeable about both and that they search for ways to incorporate best practices into their farming operations. Ever-more laws and regulations place additional requirements on growers to protect their immediate on-farm resources and environment as well as those of neighboring areas and the environment in general.

For example, Region 3, the Central Coast Regional Water Quality Control Board (CCRWQCB), requires that any grower of irrigated cropland apply for an agricultural waiver and undergo 15 hours of farm water quality education. This education includes an introduction to watersheds, on-farm water and fertilizer management, and the movement of water both on- and off-farm. In addition, farmers are required to monitor water quality through a testing program. These requirements are under review and may change in the future.

Agricultural commissioners' offices also regulate, monitor, and issue pesticide use permits according to standards set by federal, state, and local governments. All pesticide handlers, applicators and advisers must be trained and certified and must also maintain their certification through ongoing training. It is critical that you learn what regulatory requirements apply to the county in which you farm or plan to farm.

REGULATIONS, PERMITS, AND WAIVERS

Table 4.1 gives a sampling of the permits that may be required of a farm by different agencies in two localities, Santa Clara County and the San Joaquin Valley, at the date of this writing. Make sure to check with regulating agencies in your own area for detailed, current information.

RESOURCES AND REFERENCES

Avent, T. 2003. So you want to start a nursery. Timber Press, Inc., Portland, OR. Pp. 340.

California Department of Conservation, Division of Land Resource Protection. Williamson Act Program. http://www.conservation.ca.gov/dlrp/lca/Pages/Index.aspx.

California Public Utilities Commission. California Solar Initiative Program Handbook June 2010. http://www.gosolarcalifornia.ca.gov/documents/CSI_HANDBOOK.PDF.

Corum, V., M. Rosenzweig, and E. Gibson. 2001. The new farmers market: Farm-fresh ideas for producers, managers and communities. New World Publishing, Auburn, CA. Pp. 272.

DiGiacomo, G., R. King, and D. Nordquist. 2003. Building a sustainable business: A guide to developing a business plan for farms and rural businesses. Minnesota Institute for Sustainable Agriculture, St. Paul, MN, and the Sustainable Agriculture Network, Beltsville, MD. Pp. 280. http://www.sare.org/publications/business/business.pdf.

Grubinger, V. P. 1999. Sustainable vegetable production from start-up to market. Natural Resource, Agriculture, and Engineering Service (NRAES), Ithaca, NY. Pp. 268.

Intergovernmental Panel on Climate Change—The United Nations Environment Programme (UNEP GEO Team). 2002. Global environmental outlook 3, Past, present, and future perspectives. Earthscan Publications LTD., London, UK. http://www.grida.no/publications/other/geo3.

Kocher, S. D., J. M. Gerstein, and R. R. Harris. 2007. Rural roads: A construction and maintenance guide for California landowners. University of California Division of Agriculture and Natural Resources, Oakland. Publication 8262. Pp. 23. http://anrcatalog.ucdavis.edu/FarmManagement/8262.aspx.

Kurki, A., A. Hill and M. Morris. 2010. Biodiesel: The Sustainability Dimensions. ATTRA National Sustainable Agriculture Information Service. Pp. 12. http://www.attra.org/attra-pub/PDF/biodiesel_sustainable.pdf.

Maynard, D., and G. Hochmuth. 2007. Knott's handbook for vegetable growers. Fifth edition. Wiley, Hoboken, NJ. Pp. 640.

Morris, M., and A. Hill. 2006. Ethanol opportunities and questions. ATTRA National Sustainable Agriculture Information Service. Pp. 16. http://attra.ncat.org/attra-pub/PDF/ethanol.pdf.

Morris, M., and V. Lynne. 2002. Solar-powered livestock watering system. ATTRA National Sustainable Agriculture Information Service. Pp. 8. http://www.attra.ncat.org/attra-pub/PDF/solarlswater.pdf.

Olson, M. 1994. MetroFarm: The guide to growing for big profit on a small parcel of land. TS Books, Santa Cruz, CA. Pp. 252.

Rosenberg, H., et al. 2002. Ag help wanted: Guidelines for managing agricultural labor. Western Center for Risk Management Education, Spokane, WA. Pp. 242.

Ryan, D. 2009. Biodiesel: Do-it-yourself production basics. ATTRA National Sustainable Agriculture Information Service. Pp. 12. http://attra.ncat.org/attra-pub/PDF/biodiesel.pdf.

Santa Cruz County Resource Conservation District. 2004. Private roads maintenance guide for Santa Cruz County, 2nd. ed. Santa Cruz County Resource Conservation District, Capitola, CA. http://www.rcdsantacruz.org.

University of California Cost and Return Studies, Department of Agricultural and Resource Economics, Davis. http://coststudies.ucdavis.edu.

Table 4.1. Activities that may require issuance of permits from different local agencies in Santa Clara County and the San Joaquin Valley (Check with regulating agencies in your own area for more detailed information)

Issuing agency	Activity requiring a permit
Agricultural Commissioner's Office	Apiary registration Farmers markets: Certified Producer's Certificate California Organic Program Initial registration through the county. Subsequent registrations through the California Department of Food and Agriculture (CDFA). This is separate from the National Organic Program, which requires certification through accredited certifying agencies. Applying unrestricted pesticides: Operator Permit Identification number • for applying restricted pesticides: Restricted Material Permit • employer training (employees handling pesticides and/or fieldworkers; also needed for applying restricted pesticides) • Private Applicator's Certificate
Air Quality Control Board	No permits are issued by this board, but after the Agricultural Commissioner's Office issues the Verification Permit, the paperwork goes to this board. Burn Permits are issued in some counties.
California Department of Fish and Game	Depredation Permit: wild boar, raccoons, foxes, etc.
California Department of Food and Agriculture Nursery	Nursery License (paperwork available at local Agricultural Commissioner's office) Nursery Stock License (for shipping stock within California)
California Department of Food and Agriculture Seed	Registration for selling seeds
California Department of Forestry and Fire Protection; Air Pollution Districts	Burn Permits issued in some counties (Verification Permit from the Agricultural Commissioner's office required) Departments/Districts also issue Burn Permits for crop replacement, orchard pruning, range management, etc.
Central Coast Regional Water Quality Control Board	Waivers required for all irrigated commercial agricultural lands
Environmental Health	Information on requirements for diesel/petroleum fuel storage

Source: Personal communication with the Agricultural Commissioner's office, Santa Clara County.

Small Farm Profiles

The Prevedelli Family

Prevedelli Farms

Emily Ayala

Friend's Ranches

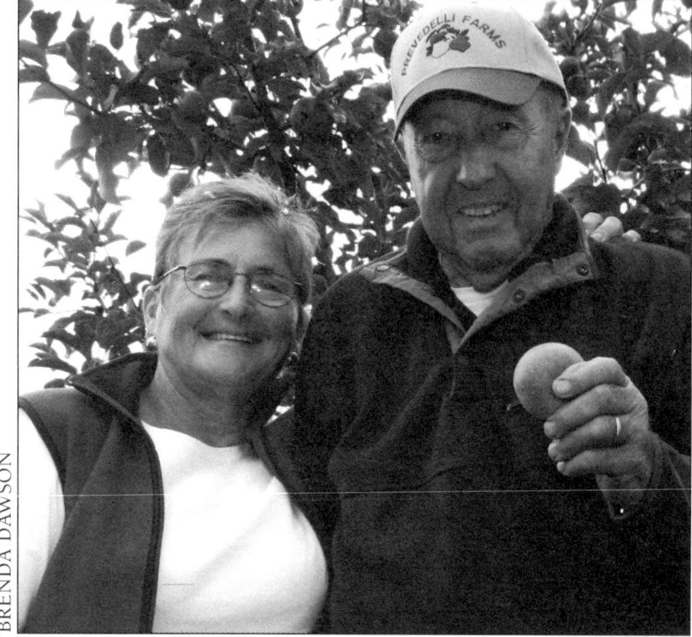

BRENDA DAWSON

SMALL FARM PROFILE:

The Prevedelli Family

Prevedelli Farms

"If you want to stay on top, always look for something new to diversify yourself," Silvia Prevedelli says. "No matter how long you are in farming, you are always learning." The Prevedelli family has been farming—and learning—for generations, first in northern Italy, where Arturo Prevedelli grew apples until 1915. After immigrating to America, he met and married another Italian immigrant, Irena Menghini, and the couple settled in Florida to farm vegetables with Arturo's brother.

In 1925 Arturo and Irena relocated to California to farm vegetables and apples on their own, eventually settling in Watsonville because of its growing climate and the availability of sufficient farmland. Two of their four children, Mary and Frank, have chosen to carry on in the family farm tradition.

On a trip to Italy to visit family, Frank met and married Silvia Torresani. The couple returned to California, where they had two children, daughter Geri and son Nick, who also eventually worked on the farm after school and during summers. Geri and Nick were the first in the family to sell produce directly to consumers: in high school, they took apples and berries to sell at a local farmers market. It was such a positive experience that the family decided to sell all of their products direct to consumers rather than through a produce broker. Today Frank, Silvia, Nick, Geri, and son-in-law Sam Lathrop actively farm together. They sell and market their fruits and vegetables through local restaurants, stores, and at 11 farmers markets, all within a 100-mile radius of the farm. Communication, building customers' trust, and sharing accurate information are the family's priorities.

"Service is very important," Silvia Prevedelli says. "We don't want them to buy an apple variety that they don't like. I always ask if they will use it for cooking or eating fresh. We pay a lot of attention to details in the sale."

The family insists that the flavor of their apples sets them apart—and Sam explains how that relates to the way they harvest their apples. "There are different ways of picking apples—either for extreme long storage or so people can eat them right away," he says. "If we were picking our apples to ship them, we would have picked them all 3 weeks ago."

Instead, they test the sugars of each variety for degrees Brix and emphasize family sampling for taste. Picking each variety at its peak and sometimes picking one variety multiple times help ensure good eating quality.

Though their apples can be stored for a short period, the family farm has expanded its year-round sales by contracting with a processor to create its own apple butter, applesauce, apple mustard, and other value-added products. The Prevedellis also sell some of their apples to a well-known cider brand. More recently, the family has further diversified the farm into other crops, including pears, blackberries, raspberries, green beans, butternut squash, and other vegetables.

Having multiple generations work together can be a challenge, but each family member brings his or her own experiences and ideas to the group. Nick spent 17 years as owner of a retail business, and Sam Lathrop's background is in construction and heavy equipment. Until recently, Geri worked part-time off the farm as an accountant, but she now works full-time in the farm business. Even Sam's teenaged son helps sell at farmers markets.

The family enjoys the social aspects of selling at farmers markets. "We have come to know people over the years. They are like family," says Geri. Sam adds, "I enjoy explaining about the different kinds of apples we sell and answering questions, which range from 'What's good today?' to 'How do you make an apple pie?'"

"While socializing is fun, the greatest thing about our work is that we are together as a family," says Geri. "We are very close, and being together is very important to us." The older generation continues to make decisions, and of course they continue to help sell at farmers markets. "We're never going to retire," Silvia Prevedelli says. "Maybe we'll slow down. We joke about getting a paycheck from them."

Together, they make decisions that take into account older traditions and newer ideas, and always the bottom line.

—Brenda Dawson

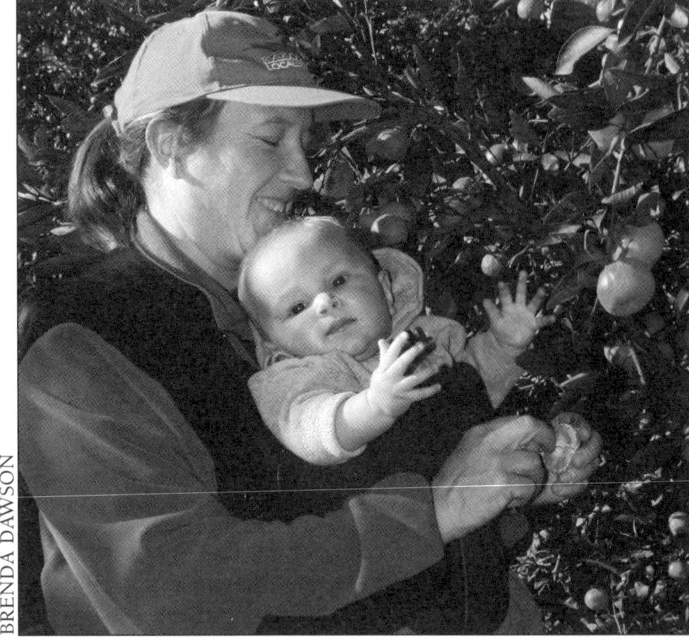
BRENDA DAWSON

SMALL FARM PROFILE:

Emily Ayala

Friend's Ranches

"I feel totally spoiled because I can take him to work," says Emily Ayala, co-owner of Friend's Ranches, with her 6-month-old son Oliver nearby. "I grew up in the packinghouse bins too."

Friend's Ranches was started by her great, great grandmother in the 1870s. When Ayala's mother and aunt inherited the business, the family decided to split the farm. Ayala's immediate family now manages about 75 acres.

The farm first grew Pixie tangerines in 1972 and is one of four family farms that founded the Ojai Pixie Growers Association to market the variety almost two decades ago. But more recently the family took about 60 percent of their land out of production in order to transition their primary crops from oranges to specialty tangerine varieties.

It takes 5 to 8 years for the new trees to reach commercial maturity, and the farm is just now beginning to approach full capacity. The transition from mainstream citrus to specialty tangerines has spread throughout much of southern California.

"If we had stuck with Navels and Valencias, we wouldn't still be here today," she says. "But whether or not the markets can support it is the big question."

Friend's Ranches sells a majority of its products at four farmers markets each week, with additional sales to wholesale markets and seasonal sales via their Web site.

Since many of the farm's tangerine varieties provide an excellent eating experience—but are more pale or bumpy than traditional varieties—Ayala says sampling is key to their success.

"People will come back for flavor even when it's ugly," she says. "Sampling is hard to do because every county has a different health inspector. You have to have all the equipment, and you need someone to talk to people. But it is really important."

Sampling is standard practice at farmers markets, but reliance on flavor can be a harder sell in the wholesale environment. Recently the family offered in-store samples at some retail stores selling their products; sales reportedly doubled.

When it comes to wholesale, Ayala says the family works to maintain relationships with wholesale brokers, friends at packinghouses, and competing farmers.

"Even though we're small potatoes, we've got to know what everybody else is doing," she says. "My five pallets aren't going to compete with their 10 carloads, but maybe they know somebody who wants my five pallets."

And vice-versa: If Friend's Ranches can't fill a customer's needs, Ayala is happy to recommend other farmers.

"Maybe I'm giving away business, but I think some time they'll send business my way too," she says.

Relationships with their employees have also proven valuable to Friend's Ranches. In addition to Ayala and her father, the farm has three full-time and two part-time employees—all members of the Estrada family who live on the ranch. Ayala's grandfather first hired the Estrada brothers in the early 1970s, and the families' professional relationship has remained strong and respectful.

"They show off the ranch to their relatives," Ayala explains. "And my parents have visited their hometown in Colima, Mexico."

Ayala manages the farm's day-to-day operations, but she also shares duties with the employees, which allows them to switch tasks and avoid too much repetition.

"I try to do everything they do—irrigation, picking," she says. "You shouldn't expect your employees to do something you wouldn't want to do."

Family and employee relationships are high on Ayala's list when it comes to defining the farm's success.

"I feel success when my employees are happy—and that includes my family," Ayala says. "When I'm able to sell fruit and get a really positive response, that just makes it all worthwhile."

—Brenda Dawson

Small Farm Handbook

THE BUSINESS SIDE

5
Enterprise Selection

KAREN KLONSKY

The selection of enterprises for the farm must be made in the context of your overall farm business strategy and its long-run goals and vision. In addition, the proposed crops should be a reasonable choice for your farm with respect to soil and climate conditions and available resources. A list of farm enterprises for California is presented in table 5.1, but it is not exhaustive, and many of the commodities listed have a large number of varieties. For example, there is a seemingly inexhaustible number of tomato varieties. This chapter's discussion focuses on crop enterprises, but the basic concepts pertain to livestock enterprises as well.

BUSINESS STRATEGY

It is critical to have a clear business strategy that will keep your farm profitable and competitive. One basic strategy is to become a low-cost producer and compete on the basis of price. Growers choosing this option often tend to lower their costs by increasing acreage, realizing economies of scale by spreading the costs of equipment and other overhead over large acreages. Growers who choose this strategy typically sell high volumes to a few large buyers, and for that reason the strategy is not generally used by small farmers. However, it can work for small farms under certain conditions: for instance, when the farm site is the ideal location for growing the crop in terms of climate or when a close proximity to market allows the grower to keep transportation costs very low. Farms taking the low-cost production approach are not highly diversified and tend to focus on one or two crops.

Another basic strategy is differentiating the farm from other operations by producing a unique product. This can be an unusual fruit or vegetable, such as olallieberries or cardoon, or unusual varieties of common crops, such as heirloom tomatoes or blood oranges. Differentiation can also take the form of extremely high quality, such as very light-colored, uniform shelled walnuts. Success using the differentiation strategy typically goes hand in hand with niche marketing—selling to a small but well-defined market such as restaurants or high-end specialty stores. The same strategy also often works in direct marketing via farmers markets or roadside stands. Using the differentiation strategy, you could either focus on production of a single commodity or run a highly diversified operation, depending on the marketing program you choose.

A third basic strategy is to be a service provider to customers. This means that you quickly and reliably respond to customer requests. Your buyers may have specific requirements with respect to packaging, quality, variety, market window, and flexibility regarding the time and amount delivered. This approach may require that you be willing to allow a customer to refuse product that does not meet his or her stringent or specified standards, or it may require that you supply your major customers first when your supplies are limited and demand outstrips supply. In some cases, the customer may request that you grow a specific crop for his or her exclusive use. In this case, the customer is actually the one doing the enterprise selection.

CONSISTENCY AND FEASIBILITY

First and foremost, each proposed enterprise should fit with the strategic approach of the business. Each enterprise under consideration must be evaluated with respect to feasibility, location of the farm, resources available, and compatibility with other enterprises. Does it build on current strengths and opportunities available to the farm? Does it expose weaknesses such as labor shortages or the potential for late spring frost?

Research the climatic, soil, and input needs of the enterprise. Consult other chapters in this handbook and the "Sources of Information" section in this chapter. If your operation is diversified, the new crop should fit into the existing crop rotation with respect to times of planting and harvesting and other labor and equipment

Table 5.1. California commodities

FRUIT AND NUT CROPS

Berries
- Blackberry
- Blueberry
- Boysenberry
- Olallieberry
- Raspberry
- Strawberry

Citrus
- Grapefruit
- Kumquat
- Lemon
- Lime
- Orange
- Tangelo
- Tangerine

Grapes
- Juice
- Raisin
- Table
- Wine

Nuts
- Almond
- Chestnut
- Macadamia
- Pecan
- Pistachio
- Walnut

Pome fruits
- Apple
- Asian pear
- Crabapple
- Pear
- Quince

Stone fruits
- Apricot
- Cherry
- Nectarine
- Peach
- Plum
- Prune

Subtropical and tropical fruits
- Avocado
- Banana
- Cherimoya
- Date
- Fig
- Guava
- Jujube
- Kiwifruit
- Loquat
- Olive
- Passion fruit
- Persimmon
- Pomegranate
- Prickly pear (cactus fruit, *tunas*)

VEGETABLE CROPS

Alliums
- Chives
- Garlic
- Leeks
- Onions
- Shallots

Brasssicas
- Arugula
- Broccoli
- Brussels sprouts
- Cabbage
- Cauliflower
- Chinese cabbage
- Collards
- Horseradish
- Kale
- Kohlrabi
- Miscellaneous leaf vegetables
- Mustard
- Radish/daikon
- Turnip
- Watercress

Chenopods
- Beets
- Chard
- Spinach

Composites
- Artichoke
- Burdock
- Cardoon
- Chicory
- Lettuce

Cucurbits
- Cantaloupe
- Cucumber
- Gourd
- Honeydew
- Squash
- Watermelon

Legumes and sprouts
- Beans, fresh market
- Peas, fresh market
- Sprouts

Other vegetables
- Jicama
- Rhubarb
- Sweet potatoes
- Yams

Solanaceous crops
- Eggplant
- Peppers
- Potatoes
- Tomatillo
- Tomatoes, fresh market

Succulent vegetable and sweet corn
- Asparagus
- Mushrooms
- Okra
- Prickly pear (cactus pads, *nopales*)
- Sweet corn

Umbels and herbs
- Basil
- Carrots
- Celery, celeriac
- Cilantro
- Fennel
- Mixed herbs
- Parsley
- Parsnip

FIELD CROPS

Grain and forage crops
- Barley
- Buckwheat
- Field corn
- Oats
- Popcorn
- Rice
- Rye
- Sudangrass
- Wheat
- Wild rice

Legumes
- Alfalfa
- Beans, dried
- Peas, dried
- Soybeans
- Vetch

Oil crops
- Jojoba
- Safflower
- Sunflower

Pasture and rangeland
- Pasture, irrigated
- Rangeland
- Woodlot

Seed crops
- Grass
- Legume
- Sunflower, confectionery
- Vegetable seed
- Tea

Sugar crops
- Cane
- Beets

LIVESTOCK, DAIRY, AND POULTRY

- Cattle (beef)
- Cattle (dairy)
- Chickens (layers)
- Chickens (meat)
- Bees (honey)
- Sheep and lambs
- Turkeys

NURSERY, GREENHOUSE, AND FORESTRY

- Cactus, aloe vera
- Christmas trees
- Edible flowers
- Greenhouse-grown container plants/ornamental
- Greenhouse-grown cut flowers or greens
- Greenhouse-grown vegetable transplants
- Outdoor-grown container plants/ornamental
- Outdoor-grown cut flowers or greens
- Outdoor-grown vegetable transplants
- Vines, canes, and other propagative matter

needs. Storage of inputs and products may also be an issue. For each crop, create a detailed calendar of operations, specifying the equipment and labor needs for each listed operation. Using this work plan you can develop a projected budget of costs.

A calendar of operations and resource use will also give you insight into whether it is reasonable for the farm to undertake the proposed enterprise, given the resources available to the farm. It will also identify additional resources (equipment, tractors, and labor) that the new enterprise may require. The timing of operations is key. By adding an enterprise, you may create a labor need during a lull time on the farm and so be able to keep a valuable employee from leaving for other work. Conversely, it may just as easily create a bottleneck when several crops have to be planted or harvested at the same time, putting a strain on available labor and equipment. A potential benefit is added income at a different time of year if your new crop also has a new harvest date. Other potential conflicts in resource use include constraints on water availability and working capital.

A sample calendar of operations for a hypothetical mixed vegetable operation with four fields is presented in figure 5.1. The calendar shows the busiest equipment months to be May and September, with major production activities in all four fields. Details could be added to the chart to show irrigation scheduling and hand hoeing in order to more completely reveal peak labor demand periods. Alternatively, separate charts could be developed for irrigation scheduling to spread out labor and water demands. From a cash flow perspective, income is generated in March, May, August, September, and October.

Estimate the prospective crop's costs of production and resource use to determine whether it is realistic for you to try to produce the crop. It may be that a crop itself would be profitable, but that the cost of transplants or drip irrigation might make it prohibitively expensive. Sample costs of production for a variety of crop and livestock operations are available online from the University of California Cooperative Extension (http://coststudies.ucdavis.edu). Each study describes a range of management practices and details the operations, input use, and costs for a hypothetical farm. A calendar of operations, monthly cash costs, and a break-even analysis are also included. An abbreviated sample study for daikon, a Chinese radish, is presented in tables 5.2 and 5.3. Table 5.2 presents the costs of production for each cultural operation, broken out by fuel, lube, and repairs, labor, materials, and custom work and rent for each operation. Table 5.3 presents the same enterprise costs but does not list the specific operations and gives greater detail about the inputs. For example, table 5.2 shows the labor cost for each operation, and table 5.3 shows the total hours of labor and the labor cost per hour. Table 5.2 gives the total cost of planting, while table 5.3 shows the number of pounds of seed and the per-pound cost for seed.

Each new enterprise you consider will require its own market research. It does you no good to produce something if you cannot find a buyer or cannot sell it at a profitable price. Take the time to become familiar with overall industry trends so you can anticipate an increase or decrease in demand for a crop and plan your enterprises accordingly. This is particularly applicable if your operation focuses on specialty items or niche markets. If possible, find out who your competitors are and see if they are in an expansion mode and how you might be positioned to compete with them.

The other aspect of marketing you should consider has to do with your customer base: will the new enterprise require that you find new customers, or will it improve your relationship with existing customers? In the first case, the new enterprise may require a significant increase in time spent marketing while in the latter case it will not.

Finally, examine the impact the new enterprise would have on your overall business vision and goals. Will the enterprise help the farm stay profitable and achieve goals related to growth measured in acres, gross income, profit, or market access? Determine how the proposed crop will contribute to long-range environmental goals related to the farm's biodiversity, soil quality, air quality, and water quality.

DIVERSIFICATION

The degree of a farm's diversification or specialization will be determined in large part by the farm business's overall business strategy. As discussed above, farms competing based on price or selling to a niche market are more likely to specialize in a few commodities or even just one. In general, specialization is based on marketing considerations and allows the farm to focus its resources on a limited range of enterprises and customers.

Enterprise diversification, on the other hand, creates opportunities for the grower to improve production, provide environmental benefits, and enhance the economic performance of the farm. From the perspective of production, for example, a good enterprise mix can serve to break insect and disease cycles, suppress some weeds,

Figure 5.1. Hypothetical crop rotation for a 14-month period.

Field number	Acres	Crop	Apr.	May	June	July	Aug.	Sept.	Oct.
1	5	Bell peppers	land preparation	plant				harvest	
		Cover crop							
		Winter squash							
2	5	Winter squash	land preparation	plant				harvest	
		Cover crop							land preparation
		Chili peppers							
3	5	Cucumber		land preparation	plant		harvest		
		Cauliflower						land preparation	plant
		Lettuce							
4	2	Sweet corn		land preparation	plant		harvest		
		Snow peas						land preparation	
Field activity by field number			Apr.	May	June	July	Aug.	Sept.	Oct.
Major production activities				All	3, 4		3, 4	All	2, 3
Income-producing fields							3, 4	1, 2	1

	Nov.	Dec.	Jan.	Feb.	Mar.	Apr.	May
	land preparation				incorporate		
						land preparation	plant
					incorporate		
						land preparation	plant
					harvest		
						land preparation	plant
			plant				harvest
	Nov.	Dec.	Jan.	Feb.	Mar.	Apr.	May
	1		4		1, 2, 3	1, 2, 3	All
					2		4

Table 5.2. Costs per acre to produce daikon in the San Joaquin Valley, 2005

	Cash and labor costs per acre ($)					
Operation	Operation time (hr/ac)	Labor cost	Fuel, lube, and repairs	Material cost	Custom/rent	Total cost per acre
CULTURAL COSTS:						
Land preparation: plow, disc, list	0.00	0	0	0	100	100
Land preparation: flatten bed tops	0.33	5	1	0	0	6
Fertilize: preplant (15-15-15)	0.09	1	0	59	0	61
Plant: seed	1.00	15	4	116	0	135
Irrigate: (water & labor)	10.50	98	0	53	0	150
Fertilize: UN32	0.00	0	0	57	0	57
Miscellaneous pickup use	5.00	75	59	0	0	134
TOTAL CULTURAL COSTS	16.92	194	65	285	100	644
HARVEST COSTS:						
Hand-pick, wash, & pack	104.00	969	0	715	0	1,684
Haul	6.00	89	75	0	0	165
TOTAL HARVEST COSTS	110.00	1,059	75	715	0	1,849
INTEREST ON OPERATING CAPITAL @ 7.65%						35
TOTAL OPERATING COSTS PER ACRE		1,252	141	1,000	100	2,529
CASH OVERHEAD COSTS:						
Liability insurance						43
Office expense						10
Land rent						300
Property taxes						5
Property insurance						4
Investment repairs						3
TOTAL CASH OVERHEAD COSTS						364
TOTAL CASH COSTS PER ACRE						2,893
NONCASH OVERHEAD (CAPITAL RECOVERY)		Cost per producing acre		Annual cost of capital recovery		Total cost per acre
Flat irrigation pipe		46		25		25
Miscellaneous field tools		100		24		24
Equipment		706		94		94
TOTAL NONCASH OVERHEAD COSTS		852		142		142
TOTAL COSTS PER ACRE						3,035

All sums are accurate, but some may appear to be in error due to rounding.

Table 5.3. Costs and returns per acre to produce daikon in the San Joaquin Valley, 2005

Cost/return	Units per acre	Unit	Price or cost per unit ($)	Value or cost per acre ($)
GROSS RETURNS:				
Daikon	650.00	box	8.00	5,200
OPERATING COSTS:				
Carton:				
Boxes, 40 lb	650.00	each	1.10	715
Seed:				
Daikon	1.00	lb	116.00	116
Custom/contract:				
Land preparation	1.00	acre	100.00	100
Fertilizer:				
15-15-15	300.00	lb	0.20	59
UN32	440.00	lb	0.13	57
Irrigation:				
Water	21.00	each	2.50	53
Labor (machine)	14.91	hrs	12.42	185
Labor (nonmachine)	114.50	hrs	9.32	1,067
Fuel: gasoline	45.82	gal	2.05	94
Fuel: diesel	2.70	gal	1.51	4
Lube				15
Machinery repair				28
Interest on operating capital @ 7.65%				35
TOTAL OPERATING COSTS PER ACRE				2,529
NET RETURNS ABOVE OPERATING COSTS				2,671
CASH OVERHEAD COSTS:				
Liability insurance				43
Office expense				10
Land rent				300
Property taxes				5
Property insurance				4
Investment repairs				3
TOTAL CASH OVERHEAD COSTS				364
TOTAL CASH COSTS PER ACRE				2,893
NONCASH OVERHEAD (CAPITAL RECOVERY)				
Flat irrigation pipe				25
Miscellaneous field tools				24
Equipment				94
TOTAL NONCASH OVERHEAD COSTS				142
TOTAL COSTS PER ACRE				3,035
NET RETURNS ABOVE TOTAL COSTS				2,165

All sums are accurate, but some may appear to be in error due to rounding.

supplement soil nutrients, improve soil structure, and conserve soil moisture. These impacts in turn can result in higher yields and higher quality, depending on the crop and production conditions. Possible environmental benefits include softening the impact of crop and livestock production on soil and water resources, curbing erosion, and increasing populations of beneficial insects and other organisms.

From an economic perspective, the most commonly cited benefit of enterprise diversification is the reduction of risk, following the adage "Don't put all your eggs in one basket." Diversification as opposed to specialization can spread out your marketing risk by expanding existing markets and opening new ones and by offsetting price swings that could affect any one commodity. Diversification can reduce production risks related to weather. It also helps reduce the grower's financial risk by spreading expenses and income more evenly over the year, and it may help keep employees on the job year-round, decreasing employee turnover and the need to train new hires.

Diversification is not without its challenges. The new enterprise may require that you develop new markets if the crop cannot be sold through existing channels. This will require market research. Planning for any new enterprise means gathering information about varietal performance, best management practices, and postharvest handling. Depending on the uniqueness of the enterprise, this information can sometimes be hard to find. Similarly, seed or transplant material may be hard to find or limited in supply. You may need to acquire additional equipment or modify existing equipment to suit the crop. You may need to expand your farm's storage capacity. Finally, the overall learning curve for running the new enterprise may be steep and may place increased demands on management time. At some point you may find that the farm has too many enterprises spreading its resources too thin and creating a situation where not enough time is devoted to any one enterprise.

SOURCES OF INFORMATION

Other farmers are always excellent sources of information and experience, and they may be willing to help you find out what has and has not worked in the past in your area. Another source of ideas and production information for small farmers is the seed catalogs from specialty seed companies. Online chat rooms and Listservs for gardeners can be useful for information on specific crop varieties and other production information. Be cautious of advice from the sales staff of propagation materials and seeds if they start recommending crops for you to grow, but at the same time recognize that a reputable company will have important production guidelines to share.

Several University of California and government Web site resources provide invaluable information. The UC Division of Agriculture and Natural Resources (ANR) Web site (http://ucanr.org) will guide you to your local UC Cooperative Extension office and show you contact and expertise information for the UC Cooperative Extension Farm Advisors in your county. The Farm Advisors' expertise can range from technical production to marketing information. Web sites for UC's Fruit and Nut Research and Information Center and Vegetable Research and Information Center, which appear below under "Resources and References," list UC experts according to crop and include important links to industry Web sites. Additional technical production information is available from UC's Sustainable Agriculture Research and Education Program (SAREP), ATTRA National Sustainable Agriculture Information Service, and the USDA Sustainable Agriculture Research and Education (SARE) program. Production manuals and pest management manuals for several crops are available from UC ANR Publications (http://anrcatalog.ucdavis.edu). The UC Integrated Pest Management (IPM) Program provides online guidelines for managing pests on numerous crops as well as information on degree-days, weather, pesticide use, water quality, and other useful areas. The UC Davis Department of Agricultural and Resource Economics maintains a Web site of its own with cost and return information for a range of crop and livestock operations, including organic production. Current market prices are available from the USDA's Agriculture Market Service for many, but not all, specialty crops.

SUMMARY

Enterprise selection is perhaps the single most important decision a farmer can make in determining the profitability of the farm business. The farmer needs to develop a basic business strategy and then select an enterprise or group of enterprises that are consistent with that strategy. Any given crop can be made to accommodate any strategy: for example, a farmer can dryland farm walnuts and be a low-cost producer, while another farmer can grow an unusual variety of walnuts and develop a niche market. A third can be a high-input producer of extremely high quality walnuts meticulously sorted for color and size and sell them to a high-end retail customer.

Part of putting together a farm business strategy is deciding how much specialization or diversification you want. Low-cost producers tend to be less diversified, focusing instead on a few commodities for which their climate or location is uniquely suited. However, different varieties of the same commodity could provide some level of diversification if you spread them out and market them over a longer period of time. Similarly, farmers who focus on differentiating their products from the same products produced by other growers tend not to be highly diversified. Yet some growers following this strategy produce a range of unusual products and service a clientele that wants unusual foods but also wants the convenience of being able to purchase a range of commodities from one source. Finally, a farming operation that offers a high level of service tends to work very closely with its customers. This may mean producing a single commodity to each customer's specifications or producing several commodities. It may also require that the grower be willing to try new commodities at the customer's request.

No matter what the level of diversification, each enterprise should be carefully and thoroughly researched before production begins. Production requirements and the corresponding equipment, labor, materials, water, and capital requirements must be identified and matched to available resources. An enterprise without a well-developed marketing program can never be successful. Market research is paramount to creating a thriving farm business.

RESOURCES AND REFERENCES

ATTRA National Sustainable Agriculture Information Service. Operated by the National Center for Appropriate Technology, with funding from USDA Rural Business-Cooperative Service. http://attra.ncat.org.

DiGiacomo, G., R. King, and D. Nordquist. 2003. Building a sustainable business: A guide to developing a business plan for farms and rural businesses. Minnesota Institute for Sustainable Agriculture, St. Paul, MN, and the Sustainable Agriculture Network, Beltsville, MD. Pp. 280. http://www.sare.org/publications/business/business.pdf.

USDA Agricultural Marketing Service, Fruit, and Vegetable Programs and Market News. http://www.ams.usda.gov/AMSv1.0/fv.

USDA Agricultural Marketing Service, Market News Service. http://www.ams.usda.gov/AMSv1.0/marketnews.

USDA National Agricultural Statistical Service (NASS), California Office. http://www.nass.usda.gov/ca.

USDA Sustainable Agriculture Research and Education (SARE). http://www.sare.org.

University of California Agriculture and Natural Resources (ANR) Publications. http://anrcatalog.ucdavis.edu.

University of California Cooperative Extension Farm Advisors. http://ucanr.org/farmadvisors.

University of California Cost and Return Studies, Department of Agricultural and Resource Economics, Davis. http://coststudies.ucdavis.edu.

University of California Fruit and Nut Information Center. http://fruitsandnuts.ucdavis.edu.

University of California Integrated Pest Management (IPM) Program. http://www.ipm.ucdavis.edu.

University of California Postharvest Technology Research and Information Center. http://postharvest.ucdavis.edu.

University of California Small Farm Program. http://www.sfp.ucdavis.edu.

University of California Sustainable Agriculture Research and Education Program (SAREP). http://www.sarep.ucdavis.edu.

University of California Vegetable Research Information Center. http://vric.ucdavis.edu.

THE BUSINESS SIDE

6
Farm and Financial Management

LAURA TOURTE, STEVEN C. BLANK, KAREN KLONSKY, AND ETAFERAHU TAKELE

The term *farm management* covers significant, essential functions in an agricultural operation. These functions include but are not limited to business planning and record keeping and the management, evaluation, and control of production, finances, and risk over time (figure 6.1). Compliance with laws and regulations at the national, state, and local levels must also be considered as part of a farm's essential functions. All of these functions are intertwined and help direct and inform a grower or farm manager's decisions. Informed business decisions are critical for making appropriate resource allocations and maintaining economic sustainability in agriculture's competitive environment.

Figure 6.1. Overview of farm management functions.

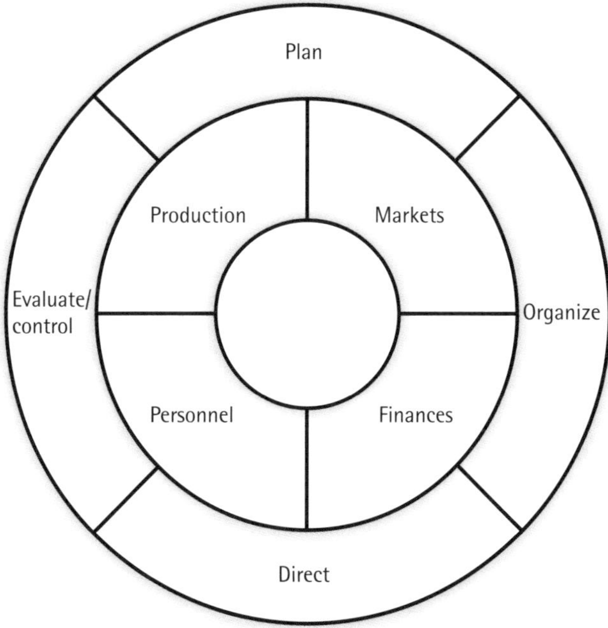

Adapted from: K. D. Olson, *Farm Management Principles and Strategies.* (Iowa State Press, 2004).

BUSINESS PLANS

Business and market plans are important foundational tools for a farm business. You can structure these plans as individual documents or combine them to form one comprehensive report. Business and market plans are multipurpose in that they define and describe a business, identify its vision, goals, and objectives, allow for the formulation of strategies and techniques to conduct the business, determine what resources are on hand and what additional resources are needed, and assess risks and challenges associated with operating the business and marketing its products. You can use business and market plans when seeking partners or funds to support a start-up or a continuing business. Finally, you can use these plans to monitor, evaluate, and guide your business over time, help you decide how and when to modify operations and decision-making processes when necessary and appropriate, and plan for future contingencies.

Developing business and market plans. Creativity, research, coordination, and organization are basic requirements when you set out to develop business and market plans. There is no single format or right way to construct the plans. Some plans are succinct and concise, while others are lengthy and elaborate. Content, organization, and clarity are noteworthy characteristics of a superior business or market plan. Table 6.1 gives you brief descriptions and suggested contents for various sections of business and market plans. Ultimately, each plan should be structured to meet the needs of the business and should contain only as much information as is necessary to fully and accurately portray the operation.

Special areas of consideration relevant to organic farming or niche markets can be included in a standard business plan, such as research that highlights organic or niche markets and trends, a grower's own experience with organic farming or alternative agricultural enterprises, familiarity with relevant laws and regulations, and special needs or requirements for a

Table 6.1. Section content of a combined business and market plan

Section	Content
Cover page	Includes pertinent information such as company name and contact information, brands, logos, Web site, and social media sites, if appropriate.
Table of contents	Serves as road map to the plan, including key sections and page numbers.
Executive summary	Presents a brief overview of the major aspects of the entire plan. It should be constructed as a stand-alone document.
Business description	Presents a concise, declarative statement describing the business, identifies its vision, goals and objectives, products and services, sources of funds, and competitive edge.
Marketplace analysis	Presents background research data and information on emerging trends and their links to the business.
Products and services	Describes products and services in more detail, including potential new product development.
Sales and marketing	Identifies and describes target market and strategies and its link to products and services. Also describes the business's strengths and weaknesses and projected income and profits.
Management	Describes administrative structure and other key personnel, with skills and experience of each.
Financial information	Presents historical data, standard accounting documents (if available), and 1 to 3 years projected budgets, including cash flow. Risk and contingency plans can also be included here.
Appendixes	Presents support documents, including résumés and graphics.

niche market or organic operation. Additional resources to help you develop plans and show you examples of completed plans are included at the end of this chapter and online at the University of California Farm Business and Market Place (http://ucce.ucdavis.edu/farmbusinessandmarketplace) under "Planning and Management."

RECORD KEEPING

Record keeping entails maintaining documentation for all aspects of an operation's business practices. A good record keeping system should be simple, accurate, and tailored to meet the farm's needs. Regular and consistent entries must be made. An accurate, well-organized record keeping system is critical in farming; it allows the grower or farm manager to track valuable production and marketing information over time. Good records help to improve management efficiency by serving as a reliable source of information, helping the grower or manager evaluate and understand what is working well on the farm and what is not, enabling the grower or manager to make adjustments to practices in response to changing conditions, and documenting the farm's financial picture and profitability. Ultimately, good records improve the farming operation's chances for success. Good records are the foundation for

- Budgets or cost and return projections
- Standard financial statements
- Labor and payroll expense tracking
- Loans and access to credit
- Tax preparation
- Systems planning for organic operations

Types of records. There are three main types of records: production records, which document information about farm production and marketing activities, financial records, which include income and expenses, increases or decreases in assets and liabilities, inventories, depreciation, payroll, and loans, and mandatory records, which include organic state registration and national certification compliance records, pesticide or chemical use records, and records to meet labor laws and regulations. Production records include information on tillage operations and crop nutrient applications, which

Table 6.2. Considerations for selecting a record keeping system

Hand-entry	Computer
Low initial cost	Higher initial cost
Easy to implement	Will require study and/or training to implement
Documentation can be time consuming	Documentation faster
Opportunities for mistakes	Greater accuracy possible
Limited analysis potential	More analysis possible

Adapted from: Establishing and Using a Farm Financial Record Keeping System, University of Tennessee Agricultural Extension Service – PB 1540, http://www.utextension.utk.edu/publications/pbfiles/pb1540.pdf.

could include equipment used, type of material and application rate, date of application, and field location (figure 6.2). Financial records document information about cash payments from the sale of crops and expenses incurred from the purchase of seed and fertilizers, for instance. Examples of mandatory records are documents that track organic product from field to fork or documents that detail health and safety trainings for workers.

Record keeping systems. Both hand-entry and computerized record keeping systems for small businesses are commercially available. Most stationery stores sell hand-entry systems and many software companies sell record keeping systems for small businesses that can be used in agriculture. Table 6.2 lists considerations for selecting a system.

There are additional considerations with respect to the selection and implementation of a record keeping system. It may be difficult or costly to find commercial software systems that meet each farmer's needs. Software developed for generic small businesses tends to record cash transactions, serving as an electronic checkbook with the ability to assign expenses and income to different cost categories. For example, cash income and expenses can be sorted for direct and indirect sales, for fresh-market versus value-added products, or by crop if there are not too many. However, these software packages may not have adequate payroll or inventory modules, the capacity to keep track of accounts receivable and payable, or the capacity to record capital investments and depreciation necessary for a small farm.

Some farmers will need to learn the fundamentals of basic record keeping, and others will need to learn how to simply use a computer before they can use the best-suited software package. Some growers need to learn both before they can even begin to learn how to use record keeping software.

BUDGETS: PROJECTING COSTS AND RETURNS

A budget is a basic economic tool that can be used to project costs and returns and analyze business decisions for a farming operation. There are several types that relate to small farm operations: whole-farm budgets, enterprise budgets, partial budgets, and capital budgets. Which type of budget you use in an analysis depends on your specific needs as a grower or farm manager. Whole-farm budgets are appropriate when an entire business is to be evaluated. Enterprise budgets are useful for determining the potential profitability of a specific crop or animal operation. Partial budgets enable you to examine the financial impact of specific changes within a crop or animal system such as changing pest management techniques or practices in a cropping system or changing to a new feeding system in an animal operation. Capital budgets are used to assess the impact of long-term business investments such as the purchase of land or equipment.

Enterprise budgets. Enterprise budgets are often the most useful analytical tools for farmers, and generally consist of two parts: the assumptions used as a basis for the analysis, and a series of tables that show estimated costs by operation and input, monthly cash costs, and net returns that are possible for a range of yields (or units) and prices. Because each farming situation varies, estimated costs are intended to help growers or farm managers understand costs, make production decisions, determine potential returns, prepare their own budgets, and evaluate production loans.

In California, enterprise budgets (cost and return studies) have been performed for a wide variety of conventional and organic crops and regions. Studies are also available for animal production systems. These studies can

Figure 6.2. Sample format for production records.

Field location				
Field description				
Total acres				
Soil type				
Irrigation type				
Water source and use				
PLANTING INFORMATION				
Planting date	Type/variety	Acres	Rate/acre	Equipment
FERTILIZERS/SOIL AMENDMENTS				
Application date	Type	Acres	Rate/acre	Equipment
PEST MANAGEMENT MATERIALS				
Application date	Type	Acres	Rate/acre	Equipment

inform and assist you as you construct your own budget. You can access them online at the UC Davis Department of Agricultural and Resource Economics outreach Web site (http:/coststudies.ucdavis.edu). Tables from a sample enterprise budget are included in Chapter 5, Enterprise Selection. Some other states have also made similar enterprise budgets available for farmers.

FINANCIAL STATEMENTS

Three standard accounting documents are generally found under the umbrella of financial statements: the balance sheet, the income statement, and the cash flow statement. Growers, lenders, and others use financial statements to help them understand and analyze a farm's financial position and its performance over time. The information in the statements provides the basis for strengthening and improving the farm business with respect to administration, organization, and operation. You can consult an accountant or a standard accounting textbook for detailed information on financial statements and how they relate to your specific farm businesses, but here is a brief introduction:

Balance sheet. A balance sheet, also called a net worth statement, provides a snapshot in time of the value of a farm's assets and its liabilities. Assets are the capital items owned by the business that have market value—for example, property or securities. To determine their value you can take their purchase price in the year of their acquisition and subtract their depreciation to date or take the price the asset would receive if sold at the present time (its current market value). Liabilities are the liens on the property including the amount of money owed to others. A balance sheet can be fairly simple, with all assets and liabilities listed and totaled in separate columns (table 6.3), or it can separate them into current, intermediate, and long-term subsections to form a more complex, detailed document. A balance sheet is considered to be a measure of a business's solvency—that is, the operation's ability to meet its financial commitments. The most helpful equation to remember with respect to the balance sheet is this:

Owner's equity = assets – liabilities

Income statement. An income statement, also called a profit and loss statement, measures the profitability of a farm over a set period of time, usually 1 year. It includes both cash and noncash income and expenses and represents profit (or loss) to the farmer for his or her capital, labor, and management. Farmers can choose to keep accounts and present the income statement in one of two ways: the cash system or the accrual system. The cash system is a simpler form of accounting (table 6.4) but is limited in the degree to which it can represent a farm's complete financial picture. The accrual system is a more detailed (and so a more time-consuming) method of accounting, but it gives a better picture of the farm's overall profitability.

Cash flow statement. A cash flow statement is perhaps the most important of the three financial statements. It documents a business's sources and uses of cash, cash in and cash out, for each month during the year (table 6.5). The data can also be presented on a quarterly or yearly basis.

Table 6.3. Sample balance sheet, in dollars (basic format)

Assets	Value ($)	Liabilities and owner equity	Value ($)
Checking/savings accounts	75,100	Accounts payable	12,300
Supplies/prepaid expenses	9,800	Current loans/credit lines	25,500
Machinery	39,400	Real estate loans	–
Real estate	–	Other liabilities	9,000
Other assets	14,600		
		Total liabilities	**46,800**
		Owner equity	**92,100**
Total assets	**138,900**	**Total liabilities and owner equity**	**138,900**

Table 6.4. Sample income statement, in dollars (basic format)

Income or expense	Value ($)	Total value ($)
Receipts		
Crop sales		
Strawberries	29,500	
Lettuce	16,200	
Mixed vegetables	12,600	
Miscellaneous sales	7,300	
Total receipts		65,600
Operating expenses		
Machinery	3,460	
Hired labor	15,800	
Seeds, fertilizers, etc.	10,000	
Interest on operating loans	850	
Office expense	900	
Rent or lease	6,000	
Miscellaneous operating expenses	2,350	
Total operating expenses		39,360
Net operating income		26,240
Fixed expenses		
Property taxes	–	
Interest on long-term loans	1,083	
Repairs and insurance on equipment	1,970	
Depreciation on assets	3,546	
Total fixed expenses		6,599
Net farm income		19,641

A cash flow statement is not a measure of profitability. Rather, it measures liquidity, showing the farm's ability to meet its credit and loan payments. Lending institutions are especially interested in a farmer's cash flow position. Historical cash flow statements become valuable tools for planning, money management, and strategic planning. They can help a farmer anticipate cash, credit, and labor needs over time.

Economic variability from year to year is common in farming, so it is best to have 3 to 5 years of financial statements on hand if you want to fully understand and evaluate a farm's financial position and performance over time. In addition to the above financial statements, farmers can perform other analyses to help them understand their farm businesses. These can include a variety of ratios and measures of profitability, solvency, liquidity, and efficiency. Formulas and explanations for these and other reports can be found in standard accounting texts. To be successful in farming, you have to make financial analysis a part of your regular business practices.

ACCESS TO CREDIT

Everyone in agricultural production needs to borrow money at some time or another. This is true for even the most profitable operations, just because of the nature of the long cash flow cycle in agriculture—the length of time from the date the first dollar is spent on inputs (for example, seeds and fertilizer), through the entire production process, to the date when the last dollar is collected from the sale of outputs from that process. The cash flow cycle for a single annual crop can last 18 to 24 months! This is partly because a farmer does not typically receive payment as soon as a crop is harvested or even immediately after the crop is sold. Many sales contracts call for payments to farmers to be stretched out over weeks or even months after the product is delivered to a buyer. This means that a farmer's cash out-flows come first and are followed much later by the cash in flows, a process that repeats with each crop and each production and marketing cycle. Farmers need access to a source of cash so they can bridge these gaps.

Tips for borrowers. A recent survey of agricultural lenders noted that when recruiting new customers some bankers target farmers with specific characteristics, including a low debt level, good business records, profit orientation, spouse involvement, continuity of management (such as is found in a family corporation), and innovative management. If you can develop a history in these areas, you can make your firm much more attractive to a lender. Listed below are some tips on how to approach a lender. The tips give additional insight into what lenders want to see in a potential borrower.

- Establish a set of short- and long-term goals for the business. This package should include a plan for achieving the goals. Recognize that the lender can be a useful partner in helping you achieve your goals.

Table 6.5. Sample cash flow statement, in dollars (basic format)

Category*	Jan.	Feb.	Mar.	Apr.	May	June	July	Aug.	Sept.	Oct.	Nov.	Dec.	Total
	$												
Income													
Farm income													
Crop 1					5,900	5,900	5,900	5,900	5,900				29,500
Crop 2								5,400	5,400	5,400			16,200
Crop 3				2,100	2,100	2,100	2,100	2,100	2,100				12,600
Other/misc.											7,300		7,300
Total farm income				**2,100**	**8,000**	**8,000**	**8,000**	**13,400**	**13,400**	**5,400**	**7,300**		**65,600**
Total nonfarm income†	–	–	–	–	–	–	–	–	–	–	–	–	0
Total income				**2,100**	**8,000**	**8,000**	**8,000**	**13,400**	**13,400**	**5,400**	**7,300**		**65,600**
Operating expenses													
Hired labor			1,200	1,200	1,200	1,200	1,200	1,200	1,200	1,200	1,200		10,800
Machinery repairs/rental		220	860	430	430	220	220	220	430		430		3,460
Farm building/repairs				438					437				875
Interest												850	850
Rent/lease									3,000			3,000	6,000
Seeds/plants	200			200						4,000			4,400
Fertilizers/lime				1,000						1,000			2,000
Chemicals				300			300						600
Custom hire													0
Gas, fuel, oil												400	400
Tax (payroll)													0
Insurance (farm)				438					437				875
Utilities													0
Supplies purchase			1,500			1,500							3,000
Management fee													0
Cooling													0
Harvest labor					1,000	1,000	1,000	1,000	1,000				5,000
Vehicle expense												200	200
Office expense	75	75	75	75	75	75	75	75	75	75	75	75	900
Total operating expenses	**275**	**295**	**3,635**	**4,081**	**2,705**	**3,995**	**2,795**	**2,495**	**6,579**	**6,275**	**1,705**	**4,525**	**39,360**
Total family living expenses†	–	–	–	–	–	–	–	–	–	–	–	–	0
Total expenses	**275**	**295**	**3,635**	**4,081**	**2,705**	**3,995**	**2,795**	**2,495**	**6,579**	**6,275**	**1,705**	**4,525**	**39,360**
Monthly balance	**-275**	**-275**	**-3,635**	**-1,981**	**5,295**	**4,005**	**5,205**	**10,905**	**6,821**	**-875**	**5,595**	**-4,525**	**26,240**

* Some categories were selected to be consistent with U.S. IRS 1040 tax preparation Schedule F, Profit or Loss from Farming.

† Nonfarm income and family living expenses were not included in this example but may be included on some statements.

- Provide the lender with an accurate balance sheet and income statement each year. If you are dealing with a new lender, provide these statements for the past 3 to 5 years, if available. As noted above, just preparing these statements will make you more aware of your farm's financial situation.

- Project the monthly cash flows for your agricultural business for the coming year and review it with the prospective lender. A borrower's cash flow projection tells the lender when the business expects to be short of cash and in need of borrowed money. The borrower can then do a better job of planning for credit needs, and that makes the lender's job that much easier.

- Maintain a good set of farm business records and share them with the lender. Lenders are impressed when a borrower can show how he or she uses such records to help make sound management decisions.

- Develop a repayment program for the loan request and present it to the lender. Include an analysis that shows the influence of a new investment on profit. Regardless of how much collateral you have, your proven ability to repay the loan is a crucial requirement of borrowing today.

- Maintain good communications with the lender. Keep the lender well informed of any developments that could prompt you to ask for additional credit or that could interfere with your making timely loan payments.

All in all, agricultural producers who want to borrow money must practice good management techniques in all vital areas of the business. It is no longer enough to have a good crop or livestock operation. A farmer can manage well in the production area and still fail because of poor management in other important areas of the business, such as marketing and finance. By demonstrating an awareness of the importance of credit management, you will improve your chances of attracting a supportive lender.

Shopping for a loan. Borrowed money has a cost, and that cost is called interest expense, the terms of which vary between lenders. That means it is important for you to shop for the best loan to fit your farm's needs. To make that shopping effort more successful, you can start by gathering the information listed in table 6.6.

Small-scale and beginning farmers may have trouble getting access to credit if they do not have cash or assets available to use as loan security. Some banks may hesitate from making small loans if they do not think the chance of a small profit is enough to offset their costs and risks in making the loan.

Despite such problems, these farmers may be able to access credit from four types of lenders. A brief description of each lender type appears below, and sample questions are given in table 6.7.

- *Farm Credit System (FCS).* FCS is a quasi-governmental agency that offers credit-related services through a network of lenders consisting of both farm credit banks and agricultural credit banks

Table 6.6. Documents a lender may request

Document	Content
Business plan*	Information about entire farm operation.
Farm business information	Resources and access to (land, water, soil type); production history; crop plans and maps, current production (including quantity and value); lease agreements; market arrangements.
Financial records	Balance sheets, income statements, cash flow statements (3 to 5 years preferred), collectible notes and accounts receivable, detailed information on outstanding loans.
Property information	Equipment and land (if owned), including descriptions and serial numbers.
Résumé	Show agricultural education and experience.

* If business plan is sufficiently comprehensive it may satisfy requests for all other documentation.

(FCB and ACB). Under the umbrella of these banks are agricultural credit associations (ACAs), which make short-, intermediate-, and long-term loans, and Federal Land Credit Associations (FLCAs), which make long-term loans. The FCS also regulates other lending entities. Notable among these are the Federal Agricultural Mortgage Corporation (Farmer Mac), which provides funds for farm mortgage loans through other lenders, such as small banks.

- *Farm Service Agency (FSA).* FSA is an agency within the United States Department of Agriculture (USDA) that often provides loans to farmers, especially small farmers, who are denied credit by other lenders. FSA provides and guarantees operating, emergency, and land purchase loans. In addition, FSA provides conservation and disaster loans.
- *Commercial banks.* Commercial banks can be anything from large institutions to small, local or community banks. The size of loans available depends on the primary source of funds: for example, funds may draw either from deposits or from capital investments.
- *Individuals and other sources.* Credit or loans from individuals or nonbank companies can take any of a number of forms, including a land contract real estate sale, an operating loan granted by an individual (for example, within a family), or a loan granted by a supplier of agricultural equipment or supplies. In this last case, the grower can only use the credit to facilitate purchases.

Table 6.7. Sample questions associated with loan inquiries and applications

Category	Question
Questions to ask a lender	Can real estate be used as security? What is the appraisal fee?
	Is there an application or commitment fee? Is the fee returned if the loan is made? Is it returned if the loan is not made?
	Are there closing costs, inspection fees, charges for documents and who pays for them (usually paid for by borrower)? Can they be made part of the loan?
	Will maintaining a bank account at the lending institution reduce the cost of the loan?
	What are the rate/term options?
Questions a lender may ask you	Are you a cosigner, endorser, or guarantor for debts incurred by others?
	Do you owe income tax?
	Are you involved in pending lawsuits?
	Are you involved in any contracts? What happens if you fall short of your contract?
	Do you have an established market for your products?
	What are the long-range plans for your operation, including changes in crops, size of operation or ownership?
	Is there potential liability in connection with violation of local, state, or federal laws or regulations?
	Most lenders ask you to fill out an environmental questionnaire asking whether there have been pollution problems, the history of chemical usage, and whether there are underground fuel tanks or chemical storage areas on the property.
	What are your sources of water? Cost of water?
	What risk management techniques are you using with respect to production, marketing, and finance?

RISK MANAGEMENT

Agriculture is one of the riskiest industries in the American economy because there is such a high degree of uncertainty regarding the potential outcomes of its production, business, and marketing activities. In this regard, risk can be taken to mean a number of different things, such as uncertain outcomes or variable outcomes over time. For a business manager such as a farmer, uncertain outcomes make planning and decision-making much more difficult. Risk, therefore, is something that a manager would want to reduce or eliminate. Farmers and ranchers must recognize their sources of risk and develop plans for managing each risk. This is critical for running a successful operation.

Risk strategies. Table 6.8 presents a partial list of major sources of risk for the small farm operation and some strategies you can use to manage that risk. The message is simple: multiple management strategies have been developed to reduce risk in farming because no single strategy can eliminate all of it. For example, production risk involves variability associated with weather, pests, and production methods, among other things. Multiple strategies are generally needed to deal with all sources of that risk.

A simple approach for developing a risk management plan involves two steps. First, identify all sources of risk faced by the business. Second, choose a strategy for managing each of those risks. The key for the producer is to become familiar with the strengths and weaknesses associated with each approach so he or she will be able to assemble the best combination of strategies to use as part of the farm's risk management plan. If there is no risk tool available to directly manage a risk exposure, look for indirect strategies that will let you accomplish the same goal.

Here is a simple example to illustrate the complexity of risk management. It involves financial risk and shows how different sources of risk can actually be related to one another. One important type of financial risk faced by farmers and ranchers—income risk—is the variability of net income levels from one year to the next. Income risk involves both production and market risks in that it includes variability in input prices, yields, and prices received (income). Because real-world risk is so complex, multiple risk management strategies and tools should be part of the management plan.

Yield risk. Until actual harvest there is always uncertainty about the yield, which makes forecasts of sales revenues uncertain—and this is called called *yield risk*. A grower can reduce the financial effects of yield risk by purchasing crop insurance from either a private insurance company or the Federal Crop Insurance Corporation (FCIC). USDA's Farm Service Agency offices can direct you to insurers who sell federal policies. Crop insurance pays the farmers when the farm's average yield for a year falls below some insured level, which is based on the producer's historical yield averages.

Price-received risk. The price a grower will receive when selling an agricultural product is the most uncertain component when forecasting sales revenue. To reduce the risk associated with this uncertainty, some producers negotiate forward contracts with buyers, actually setting the price to be received by the producer, often done before the production process has begun. Forward contracts may cover a single year or multiple years. Forward contracts have not been developed or made available for all crops, and may be challenging for small farm operations to use. Another tool is the USDA Agricultural Market Service (AMS), which compiles current and historical market price information that can be used to estimate potential sale prices for farmers.

Input price risk. After calculating sales revenue, you also need to estimate input costs before you can estimate gross income. The prices you will have to pay for inputs often are not known until you actually make the purchases. In some cases, you may be able to reduce this risk by establishing forward contracts for inputs. You may also be able to negotiate lower unit prices for some inputs if you are able to purchase the inputs in sufficiently high volumes.

Even when you try to manage risk with regard to yields and prices, you still may not be able to reduce income risk enough to satisfy a lender or to make yourself comfortable with the degree of risk exposure you face. For this reason, most producers in California use crop diversification and differentiation of product as income risk management tools. By diversifying their production efforts into multiple crop or livestock enterprises, producers can reduce the risk of being victim of a down market for any single commodity. In other words, if a farmer sells multiple products, a low price for one of the crops in a given year can be offset by a good price for another crop that same year. This

Table 6.8. Sources of risk and strategies to manage risk in farming and ranching

Source of risk	Strategies to manage risk
Production	Select enterprises with low production risks. Diversify business and products. Maintain flexibility with input cost arrangements. Use production practices that reduce risk (e.g., irrigation). Purchase crop insurance to insure yield and income. Farm in more than one location. Lease land or equipment; hire custom operators. Invest in extra machine capacity. Maintain cash and other resource reserves. Obtain additional production information.
Market	Select enterprises with high consumer demand. Diversify business and/or differentiate products. Maintain product and harvest cost flexibility. Spread product sales over time. Negotiate contracts or other market arrangements. Obtain additional market outlook information.
Financial	Off-farm employment and other types of off-farm income. Maintain financial or credit reserves. Negotiate longer loan repayment periods. Maintain safe level of debt. Develop land lease strategies. Incorporate to limit risk. Obtain property insurance to mitigate possible losses. Obtain additional accounting training and information.
Technology	Maintain flexibility and update as appropriate. Keep informed of new developments.
Legal	Comply with requisite laws and regulations; document compliance. Keep informed of new laws and regulations. Maintain adequate insurance program.
Human	Develop and maintain strong personnel policies and procedures. Update and maintain appropriate health and safety trainings. Plan for loss of management and/or employees. Maintain insurance program. Plan for estate transfers.

TAX AGENCIES FOR AGRICULTURE

Local County. The county tax assessor determines the fair market value and the county auditor computes the taxes on land, buildings, and equipment. Contact information is listed in the "County Government Offices" section of local telephone books and can also be found on the Internet.

State of California. To find contact information for state agencies and boards with relevance to agriculture, look in the "State Government Offices" section of your local telephone book or on the Internet. These include

- *Franchise Tax Board,* which has information on California income and corporation taxes and distributes tax forms.
  Telephone: (800) 852-5711
  Web site: http://www.ftb.ca.gov

- *Board of Equalization,* which administers fees and taxes for various divisions.
  Telephone: (800) 400-7115 (statewide)
  Local office locations and telephone numbers may also be available. Check your local telephone book.
  Web sites include
  - *Sales Tax and Use Tax.* Sales tax permits for agricultural products. http://www.boe.ca.gov/sutax/sutprograms.htm
  - *Environmental Fees Section.* Includes hazardous waste. http://www.boe.ca.gov/sptaxprog/spenvirofees.htm
  - *Fuel Taxes Section.* Includes underground storage tank maintenance fees. http://www.boe.ca.gov/sptaxprog/spfuel.htm
  - *Timber Yield Taxes Section.* http://www.boe.ca.gov/proptaxes/timbertax.htm

- *State Controller's Gasoline Tax Refund Division,* which refunds road-related taxes that have been paid on gasoline that growers use in off-road farm vehicles.
  Telephone: (916) 445-4868
  E-mail: gtr.sco.ca.gov
  http://www.sco.ca.gov/ardtax_gas_tax.html

- *Employment Development Department,* which has information on unemployment insurance tax, disability insurance, income tax, and programs that provide tax credits to employers.
  Unemployment insurance telephone: (800) 300-5616
  Disability insurance telephone: (800) 480-3287
  Local offices and telephone numbers may also be available. Check local telephone books.
  http://www.edd.ca.gov/

Federal. *The Internal Revenue Service (IRS)* collects federal income tax, social security tax, and unemployment tax.
Telephone: (800) 829-4933 (nationwide).
Local telephone numbers can also be found on the IRS Web site: http://www.irs.gov/localcontacts

Several publications are available from IRS to help growers and farm managers prepare their federal tax returns. Titles and publication numbers include

- *Agricultural Employer's Tax Guide,* Publication 51 (Circular A)
- *Farmer's Tax Guide,* Publication 225
- *Tax Guide for Small Business,* Publication 334
- *Starting a Business and Keeping Records,* Publication 583

Additional business information, forms, and publications can be found at http://www.irs.gov/businesses.

averaging effect reduces the producer's fluctuation in total income and so reduces overall risk. This strategy can be very effective for a small-scale farmer and is highly recommended. However, too much diversification can create its own problems by requiring mastery of more marketing and production skills and possible requirements for specialized equipment and labor.

TAXES

Taxes and tax law both change over time, so no general guide such as this handbook can or should offer a list of details. Due to the complexity of the tax issues facing agricultural producers and the serious consequences of making any error regarding tax payments, the authors recommend that you seek assistance from a professional tax adviser. The best source of assistance is most often a Certified Public Accountant (CPA) with agricultural experience who is up-to-date on tax law. To get an idea of the complexity of tax issues in agriculture, take a look at the sidebar "Tax Agencies for Agriculture" for a list of contacts for county, state, and federal agencies that administer and collect taxes from farm operators. General and useful information is also available at http://www.ruraltax.org.

RESOURCES AND REFERENCES

Multiple resources associated with farm management and marketing are located at the following one-stop Web site: http://ucce.ucdavis.edu/farmbusinessandmarketplace.

DiGiacomo, G., R. King and D. Nordquist. 2003. Building a sustainable business: A guide to developing a business plan for farms and rural businesses. Minnesota Institute for Sustainable Agriculture, St. Paul, MN, and the Sustainable Agriculture Network, Beltsville, MD. Pp. 280. http://www.sare.org/publications/business/business.pdf

Grubinger, V. P. 1999. Sustainable vegetable production from start-up to market. Natural Resource, Agriculture, and Engineering Service. (NRAES), Ithaca, NY. Pp. 268.

Olson, K. D. 2004. Farm management principles and strategies. Iowa State Press, Ames, IA. Pp. 429.

Pisoni, M. E., and G. B. White. 2002. Writing a business plan: A guide for small premium wineries. Cornell University. EB 2002–06 Department of Applied Economics and Management, Ithaca, NY. Pp. 35.

Richards, S. 2002. Getting started in farming? Five keys to success. Cornell University. Department of Applied Economics and Management, Ithaca, NY. Smart Marketing Series. Pp. 4.

Schlough, C. 2001. Knowing your market—The most challenging part of a business plan. Cornell University. Department of Applied Economics and Management, Ithaca, NY. Smart Marketing Series. Pp. 4.

Uva, W. L. 1999. Travel the road to success with a marketing plan. Cornell University. Department of Agricultural, Resource and Managerial Economics, Ithaca, NY. Smart Marketing Series. Pp. 4.

White, G. B., and W. L. Uva. 2000. Developing a strategic marketing plan for horticultural firms. Cornell University. EB 2000–01. Department of Agricultural, Resource, and Managerial Economics, Ithaca, NY. Pp. 26.

THE BUSINESS SIDE

7
Marketing and Product Sales

SHERMAIN D. HARDESTY

When you first saw the word marketing in the table of contents, you probably thought about advertising or sales. Actually, marketing encompasses a broader range of activities, starting with research, target market identification, and planning, and moving into product selection and development, package design and labeling, pricing, advertising, sampling, public relations, promotional events, selling, distribution, and evaluation. This chapter addresses all of these activities, with greater emphasis on some than on others. As you read, be aware of the power of marketing: your marketing decisions will affect how much profit you earn!

MARKET RESEARCH: FIND OPPORTUNITIES BY UNDERSTANDING CUSTOMERS AND COMPETITORS

Small growers spend many hours producing their crops, but many do not earn a fair return—earnings that are high enough to compensate them for their management efforts and personal labor, after covering production costs, cash overhead expenses, depreciation on machinery, equipment and structures, and the rental value of their land. Understanding a few key marketing concepts can help you improve your profitability and earn a fair return. It will simplify things if we discuss the marketplace first.

As a grower, you face many decisions about what to produce and how, when, and where to market your products. Because the marketplace can be very confusing, it helps to divide it into two sectors—customers and competitors—and then set about to understand each sector. This market research effort will help you identify meaningful opportunities. You may sell direct to consumers or to grocery stores or restaurants, but consumers are still your ultimate customers. Customers, whether current or potential, are often identified by demographic characteristics as well as their attitudes and life-style, such as busy working mothers, environmentalists, health-conscious seniors, and foodies seeking the latest, unusual products. Talk to customers to learn about their needs and identify the kinds of products that you could supply them with. Ask them what kinds of products they are most interested in buying and why. Don't forget to talk to retailers and restaurants—but remember to focus those efforts. A manager at a big-box discounter is not an appropriate information source if you are trying to sell specialty cheeses! By reading newspapers (particularly the weekly food section) for the major metropolitan areas closest to you, will also get an idea about emerging consumer and food trends. Even if you are not selling to restaurants, it is important to monitor hot items at restaurants because most food trends are introduced there.

The local food movement is opening new doors for small producers. Consumers are buying more locally produced foods because they want fresher, tastier food, they want to have a human face they can associate with the food they eat, and they want to support more sustainable agriculture production and a diverse, stable regional food economy.

Don't forget about your competitors, either. Look around and see who is successful. While you don't just want to copy them, you may be able to get good ideas from some of their tactics, such as packaging or signage. Most of all, you want to find ways to distinguish yourself from the competition. Look at your competitors' products and displays at various venues, such as farmers markets, groceries, and specialty stores—and don't forget to check out their Web site on the Internet.

UNDERSTANDING MARKETING

A basic understanding of a few key marketing concepts will help you as you make critical decisions regarding your crop mix, customer selection, selling and promotion tactics, and pricing. In the first edition of this handbook, growers were described as being for the most part production-oriented because they tended to produce the crops first and then think about how to get people to eat more of what they had produced. To be a successful marketer, a grower needs to be market-oriented: first identifying the customers' needs and desires and then producing products to address those needs and desires.

The key to successful marketing is to get enough customers to pay a price that generates a fair return to you, the producer. As a small-scale farmer, you have higher per-unit costs than large-scale farmers do. Because of this, you cannot compete on a price basis in mass markets; instead, you need to pursue a niche market strategy. If your production is very small in scale, your niche market strategy may be very narrowly focused—say, selling pumpkins to school children and their parents who visit your farm. As a medium-scale producer, your niche market strategy would address more niches, such as selling specialty vegetables at ethnic grocery stores, urban farmers markets with a large ethnic customer base, and ethnic restaurants close to the farmers markets.

By talking to consumers and other customers, you can identify your target market. Targeted marketing is the opposite of mass marketing, which is the typical production-oriented approach that aims at everyone. Instead, you will be focusing your products and marketing efforts on your identified target market. By focusing in this way, you reduce the number of choices you have to deal with considerably. For instance, if your target market is the educated environmentalist, you can immediately rule out selling your produce at flea markets and concentrate your marketing on farmers markets and specialty markets in college towns. Some producers have two sets of target markets; they sell their perfect produce to upscale regional grocers and their cosmetically challenged (but still perfectly edible) produce at a farmers market in a lower-income area.

Next, consider the four P's of the marketing mix: product, place, promotion, and price. All four factors require your careful consideration if you want to maximize your effectiveness in your target market.

Product includes the actual crops and their special characteristics, such as organic production, heirloom varieties, or other specific, unusual varieties. Many small farmers grow a broad range of crops as a way to protect against crop failures. Many also seek to maintain a presence in the marketplace throughout the marketing season by planting different varieties and staggering their plantings so they can get their produce in the market as early and as late in the year as possible. If you have a hot product, you can try to use it as leverage to market your less popular crops; for instance, when you negotiate with a grocery store to buy your popular white and yellow corn, you can tell them that you won't sell it to them unless they also buy your cucumbers. Product also includes your brand or farm name and your packaging. Pay attention to what works best in your target market: a brand name or packaging approach that is effective in one market could be unappealing in another market. For example, wrappers in subdued colors and made from recycled paper could be very desirable to a target market of environmentalists, but the same wrappers are unlikely to appeal to upscale buyers of gourmet foods.

Place refers to the distribution channels you use to reach consumers. These could include direct sales to consumers through farmers markets, farm stands, and community supported agriculture (CSA), direct sales to grocery stores, ethnic specialty markets, and restaurants, or indirect sales through wholesale markets, distributors, and packer-shippers. Geographic location is also part of place; you may need to drive the extra distance to sell at a farmers market in a large urban area with a high ethnic population or to sell to upscale regional grocers in college towns. The choice depends on your target market. You can read more about the different channels later in this chapter.

Promotion is also important, since you can't assume that consumers will be aware of your products without it. Clearly, you won't be advertising your products on network TV; instead, you need to consider other ways to raise consumers' awareness, such as displaying your logo, using appropriate signage for your farmers market stall, sending out an e-mail newsletter, providing press releases to local newspapers, and participating in special events. Again, how you promote depends on your target market. Benefit tastings in retirement communities can be a good fit if you are targeting health-conscious seniors, but not if you are targeting active environmentalists. There is more information about promotion tactics in the sidebar, "Promotion and Communication Tactics," later in this chapter.

Price needs to meet two conditions: it must be right for your target market and it must be high enough to make you a profit. Know your costs of production, including fixed costs such as the payments on your farmland, property taxes, insurance, and equipment depreciation. Monitor your competitors' prices. Keep your prices steady so your customers will know what to expect, but don't be reluctant to raise your prices if you have a sharp increase in your costs and market conditions permit. If you never have customers walk away because your prices are too high, then you aren't charging enough! Producers often offer volume discounts by prepackaging their produce in larger quantities.

There aren't many sources available for pricing information. Always make note of your competitors' prices wherever you are selling. A limited amount of data on wholesale prices for organic produce is available from United States Department of Agriculture (USDA) Agricultural Marketing Service (AMS) (http://marketnews.usda.gov/portal/fv). Terminal markets in San Francisco and Boston provide data for some organic produce, which is reported to the USDA Economic Research Service (ERS) (http://www.ers.usda.gov/data/organicprices/). Rodale Publishers' New Farm Organic Price Report (http://www.rodaleinstitute.org/Organic-Price-Report) provides weekly comparisons of terminal market prices and other wholesale prices as well as selected large-scale retail prices for organic and conventional produce and grains. The report includes three West Coast markets: Los Angeles, San Francisco, and Seattle.

Product, place, promotion, and price all contribute to how you can best market your product. The combined result of the decisions you make related to the four P's is positioning. This concept has to do with the space the product occupies in consumers' view of the world. For example, you can position your products in one way by emphasizing that they are produced with highly environmentally sound practices or in another way by emphasizing that they are a great find for bargain hunters. The result of positioning is differentiation, and effective differentiation makes your products distinct and makes them appear to be a better fit for your target market than your competitors' are. Differentiation helps your customers know what makes your products unique, and this gives them a reason for choosing your products over other similar products in the marketplace. For example, your soils and cool night breezes may build just enough acid to make your peaches more flavorful than those of your competitors.

Once you get used to managing your marketing mix, your differentiation will become very obvious and your marketing decisions will be very easy to make. The first key decision area we will discuss has to do with marketing channels.

MARKETING CHANNELS

When deciding where to sell your products, you have many options to choose from. Each of the channels listed below has its own pros, cons, and special requirements. Your marketing channel choices should be part of a well-thought-out marketing plan that is based on your personal goals, market trends, target market, relevant regulations, and your own personal circumstances.

Where and how you sell your products should be consistent with your positioning and should strengthen your differentiation. To succeed in some marketing channels you may have to meet special requirements; for example, if you decide to direct market your products, having an outgoing personality (or hiring help with that quality) will be a great benefit. If your farm is at the edge of a large urban area, a farm stand or u-pick operation could be a likely marketing option.

As you read through the options listed below, consider whether you may want to sell through multiple channels. With multiple channels, you can cross-promote your different operations, telling your customers at farmers markets about your CSA or U-pick operations. Many successful small producers find that multiple channels give them the flexibility to market products of varying levels of ripeness and grade. For example, growers can sell less-mature fruit through wholesalers and their ripest fruit through farmers markets.

Many successful small farmers started out by selling at farmers markets where they established their identity and developed relationships with the buyers for restaurants and grocery stores who became their customers. Many also combine sales of food products with other revenue-generating activities, such as production and sales of homeware items and firewood from pear tree prunings, pear-inspired photographs, cards, stories, and poetry.

Direct to consumers. When you sell direct to consumers, you get the opportunity to discuss your products with potential customers, receive immediate feedback from customers, and connect your identity with your products. USDA reports that farmers received about $.25 to $.30 of every dollar consumers paid retailers for fresh fruits and vegetables over most of the past decade (USDA ERS 2009). Direct marketing can earn you full retail ($1.00 of every dollar), but your selling costs (such as transportation and labor) can be considerably higher than if you were to sell through a packer-shipper or distributor. In California, produce that is sold direct by the producer to consumers is exempt from the state's

PROMOTION AND COMMUNICATION TACTICS

Here are some tips on how to create consumer interest in your products. Remember to think about your positioning as you develop your promotion program!

- **Public relations.** Coverage by a local newspaper or radio or television station can significantly boost your sales. When you write a press release, emphasize something new, such as an award, new crops, or an upcoming event on the farm. Invite a local newspaper's food editor over for lunch or take lunch to a local radio or television station and let them rave over your tree-ripened peaches or grass-fed beef.

- **Signage.** Consider your business signage to be a form of advertising. Make it attractive and put in enough of your story to let passersby know why your products are different. Signs should have a logo or consistent, identifiable graphic look. Use similar signage at your farm, farmers markets, and events so customers will get repeated exposure to the same message.

- **Web site.** A well-designed, simple Web site can be a great tool for building customer awareness—both with businesses and individual consumers. Make sure to tell your story online. Keep your material fresh by regularly updating product availability information and recipes. Be shameless in putting your Web address on everything: including labels, business cards, T-shirts, and caps.

- **Direct mail.** Whether you mail out postcard updates or newsletters or send customers e-mail reports, make sure to collect contact information from your customers and farm visitors. Ask loyal customers to provide contact information for their friends who are interested in locally grown food. If you participate in a benefit event, ask the organizers for their attendees' contact information and add them to your list.

- **Packaging.** Whether you use ready-made containers or just simply wrap your produce with a rubber band or twist tie, consider your packaging to be your ever-present salesperson. Don't skimp on packaging design; hire a professional designer who understands the food business. Make sure you include your URL on your labels and bands.

- **Business cards, brochures, stationery.** These items are your identity package and need to tell your story. They should have a similar look and should reinforce your unique marketing position (it's probably not wise to use fluorescent orange paper with neon green lettering if you are trying to differentiate your heirloom tomatoes). Whenever you meet with potential customers, give them your business card and a brochure. Make sure your URL is included on all of your materials. Display your business card and brochures when you participate in events.

- **Partnering.** Stretch your limited marketing resources by partnering with local businesses. If you are selling salad greens at a farmers market, provide samples of your local bakery's croutons with your greens. If you are an apple producer with wineries nearby, ask a local winery to serve your sliced apples in their tasting room. Just make sure that the winery displays your brochure next to the apples and that you pick partners who are consistent with your market positioning.

pack and grade standards. You may also be exempt from marketing order assessments if your total sales of a particular commodity are less than the specified minimum.

Farmers markets. The farmers market is a common entry point for beginning farmers. You have face-to-face contact with consumers, potentially very beneficial for developing contacts with chefs and possible CSA subscribers. Unlike most other channels, you can sell small volumes at a farmers market. It's a good idea to offer samples of your product; just make sure to check with the market manager about sampling rules. While the consumer is sampling your product, you can share information that differentiates your product.

Make an effort to get to know your repeat customers and ask them for feedback about what they bought from you last week. Design your stand thoughtfully to reinforce your differentiation. You can also benefit from interaction with more-established farmers and try out new sales approaches at limited cost. Travel times to farmers markets can be significant, and it is often difficult to find good hired help if you cannot staff your own stall. Some markets are seasonal; others operate year-round. If you are trying to get into one of the more popular markets, expect to be put on a waiting list unless you have an unusual product. Markets allow you to purchase multiple spaces, but usually they group them in 10-foot increments. Some markets charge a flat stall fee ranging between $10 and $30 per 10-foot increment. Other markets assess vendors a market fee based on the vendors' gross sales, usually between 4 and 7 percent, but they may also have a required minimum fee.

Fairs and flea markets. Fairs and flea markets are similar to farmers markets, but they are not held as regularly. However, you can still develop a regular customer base at a flea market. Consumers will be looking for bargains, so make them feel like they get a good deal by adding some extra product to their box or bag— but make sure to build the cost of this premium into your pricing. Expect to haggle over prices with some consumers. Fairs and flea markets can be a great way for you to sell larger volumes than you can sell at farmers markets, but your prices will have to be lower. Make sure to evaluate your positioning carefully before you decide to sell at fairs or flea markets. If customers are introduced to your products at a fair or flea market, they could be resistant to paying higher prices at your regular sales venues.

Community supported agriculture. A CSA is a subscription program involving an ongoing relationship between you and your customers. Subscribers pay you in advance to receive a box of products, usually once a week. When CSAs were first introduced, subscribers paid the farmer an annual fee to cover the farm's production costs and received a weekly share of the harvest during the growing season. Some such arrangements still exist, but many CSAs in California now have quarterly subscriptions; others bill their subscribers for each box delivered and allow them to cancel their subscription at any time. Compared to selling through distributors, a CSA can speed up your cash flow significantly and provide you with much-needed operating capital. Most CSAs use established pickup sites for their produce boxes, and a few deliver direct to the customer's home. The boxes usually include recipes for the products and a newsletter from the grower. Subscribers usually pay full retail for the products. Some growers exchange products with other growers to provide a broader selection in their boxes. A grower with a mature CSA program can generate a significant, regular revenue stream, but they have to manage their product selection and offer appealing benefits (such as farm visits) to keep their annual member retention rates at a sustainable level (about 67%).

Farm stands. A farm stand will also generate full retail-priced sales for a producer. Although you do not have to transport your products as far as you would for a farmers market, you will incur other costs to operate your stand, such as labor and utilities (refrigeration power, for instance). Stand construction costs can vary from almost nothing for a temporary stand to well over $100,000 for a permanent, more elaborate store. Remember that location is critical; the site needs to have relatively high traffic. You will need to provide parking and safe access from the road. Attractive signage to identify your stand is also important. You may need to advertise in local newspapers and travel guides. To broaden your product selection, you may want to resell products from other producers. Regular hours and friendly service will help bring back customers.

Many produce stands also offer processed products (more information on this later in this chapter) such as jams, dried fruit, and baked goods that are often made from produce that gets too ripe or is otherwise unsellable. You will need to carefully review local government regulations when you investigate setting up a roadside stand—and be prepared to pay for permits. Regulations can vary widely by city and county. New state regulations allow farmers to sell some bottled water, sodas, and other nonlocal foods as long as the selling space for these nonlocal products does not exceed 50 square feet. You will find more information on state regulations later in this chapter and online in the volume 2, 2009, edition of the UC Small Farm Program newsletter (http://www.sfp.ucdavis.edu/pubs/SFNews/200902news.pdf).

U-pick operations. U-pick operations are increasing along with the popularity of agricultural tourism. Consumers want fresh produce and enjoy driving out to the country to visit with a farmer and pick their own fruits and vegetables. As a grower, you have no harvest costs (though you do need to provide picking boxes or baskets), and it is not critical that you have a wide crop selection. As with roadside stands, location is critical and your weekend hours can be long. The visitors can also be an intrusion to your family life and your farming activities. There are significant liability issues, so make sure to review your plans with your insurance agent to ensure that you are adequately protected. You will also need to contact local government agencies to make sure you have appropriate zoning, parking, and other facilities. U-pick operations often have a stand where visitors can also buy harvested produce and processed products.

Rent-a-tree. One specialized form of U-pick, the rent-a-tree program, is becoming popular for fruits, nuts, and olives. As a grower, you have to take care of the same liability and government regulation issues as for any U-pick operation. Unlike other U-pick programs, you share the crop risk with your rent-a-tree subscribers and you have their advance payments to bolster your cash flow. You should plan to establish a long-term relationship with rent-a-tree customers. Most rent-a-tree programs require that the subscriber participate in the harvest, while others deliver the harvested (and processed, in the case of olives) product to the subscriber.

Mail-order, Internet. Mail-order and Internet sales can be attractive if you have a highly differentiated product. You may have started at farmers markets and developed a loyal customer base—perhaps even among tourists. Make sure to let your farmers market customers know that you also have a mail-order program. Your products need to have a high price tag relative to their weight, since shipping costs can be significant.

With the Internet, you can promote your products through your Web site. You will need to invest in a well-designed Web site and coordinated graphics for your packaging and sales materials. Depending on how you set it up, your customers can place their orders through your Web site, by fax, or by phone. Make sure you get contact information for your mail-order customers (and gift recipients) so you will be able to send them reminders when your harvest time is approaching. Your farm can be in a relatively remote location, as long as you have good access to delivery services. This type of direct marketing does not require that you have great people skills, but you do need to be well organized to run an efficient mail-order business. Since your products are likely to be highly perishable, you will need to invest in packaging equipment and have ways to monitor the quality of the shipped product. If you start your mail-order program with only unprocessed products, it's a good idea to consider expanding into processed products to extend your marketing season, perhaps even to year-round. This will keep your customers thinking about you.

Direct Sales: Other Categories

The rest of the marketing channels described here involve direct sales to businesses rather than to consumers, but you still need to be aware of trends and understand who their ultimate customers are and what their needs are.

Direct to restaurants. Selling direct to restaurants has become a popular practice for small producers. Chefs are buying from local farmers and ranchers because of the quality, freshness, and uniqueness of the food, their relationships with the producers, and customer requests for local products. Be aware, though, that you'll also encounter numerous challenges when selling to restaurants. Most chefs want product consistency and delivery reliability—they want you to deliver the product in the quantities desired and at the times promised. They may feel hampered by the limited availability and variety of your product mix. Many small producers make their chef contacts at a farmers market. If you are initiating contact with a restaurant, do some homework first by getting to know the restaurant—maybe even eat there—and then ask to speak with the owner or chef.

When you meet with the person who makes the buying decisions, let them taste your products and be sure that you tell your story—what makes your products different. Build a relationship by staying in touch—check on their satisfaction with your products and inquire about their unmet needs. Some restaurants will continue to pick up their orders at the farmers market while others will require delivery directly to their restaurant. It is important to fax or e-mail chefs a weekly fresh sheet of what you have available. Also, you will need to be vigilant in collecting payments for your restaurant accounts (preferably at the time of delivery); unfortunately, restaurants have a relatively high rate of failure, and that means you run the risk of getting stuck with an unpaid bill. Recent California legislation (AB 2168) permits farmers to sell directly to chefs and nonprofit organizations at farmers markets, farm stands, and field retail stands, while keeping the sold product exempt from wholesale size and pack regulations. However, the grower must provide the buyer with a bill of sale or similar documentation listing the grower's name and address as well as the type and quantity of product purchased.

Direct to grocers and specialty markets. Opportunities to sell direct to grocers and specialty markets are increasing as many regional grocery stores focus on locally produced foods to differentiate themselves from national chains. You should also look into selling to natural foods markets, grocery cooperatives, and ethnic food markets; again, the specific markets and locations you choose will depend on your positioning. Selling to these markets is similar to selling to restaurants, but with higher sales volumes. If you are initiating contact with a store, do some homework first to get to know the store—maybe even shop there—and then ask to speak with the produce manager. Fax or e-mail a weekly fresh sheet to the produce manager and send a newsletter similar to the one you would send to CSA customers. Make sure that your invoices are clear and that you understand the store's payment practices.

Direct to institutions. The local food movement is creating opportunities for growers to sell direct to institutions—hospitals, schools, colleges, retirement homes, and even jails and prisons. These institutions usually operate on a restricted budget, but some will save on other items in order to be able to buy locally produced foods. Some operate their own foodservice program while others contract out to a foodservice operator. Again, do your homework first to get to know the institution you plan to approach, and then ask to speak with the chef. Although the chef may not do the buying, he or she is most likely to initiate local buying programs.

As with restaurant chefs, let the institution chef taste your products and make sure to tell your story—what makes your products different. Ask questions that will help you understand their specific needs, such as preferred fruit size, packaging, and delivery scheduling. Build a relationship by staying in touch—check on whether they are satisfied with your products and inquire about their unmet needs. Because of the large number of meals served by an institution, you will need to be very conscientious about your delivery volumes and times; if you are having difficulties, let your contact know so that he or she can make alternative arrangements. Institutions often have special requirements of their suppliers: don't be surprised if you are required to have product liability insurance, vehicle liability insurance, and an inspection of your farm.

Packer-Shippers

Combination packer-shippers consolidate deliveries from numerous growers along with their own crops. Most packer-shippers specialize in a particular type of produce, such as stone fruit or leafy greens. They grade, pack, cool, and market produce to a customer base made up of distributors, exporters, retailers, and foodservice operations. They often have direct seasonal contracts with grocery and foodservice chains for multiple products, and they need to consolidate the production of numerous growers in order to meet the volume requirements associated with these large customers. If you don't have cooling and storage facilities and don't want to direct market, you will probably need to sell through a packer-shipper. Many packer-shippers have contracts with growers to handle and market certain crops. Usually, these contracts do not specify a price; rather, they are agreements to sell the product at prevailing market prices. Some packer-shippers buy produce directly from growers, but most pay growers a net price after they deduct their sorting, grading, packaging, cooling, shipping, and marketing expenses from the market price received.

Wholesalers and Distributors

Wholesalers and distributors are the traditional middlemen in the produce distribution system. They buy from growers and shippers, transport the produce to a central location, and resell it in smaller quantities to retailers, institutions, restaurants, and regional distributors. Some will pick the product up from multiple farms and shipping points on a route, deliver it to a centralized distribution center, and ship it immediately to another region. If you want to work with a wholesaler or distributor, you need to make sure that you can meet their volume requirements. In California and many other states, the grower needs to meet pack and grade requirements for the specific crop, such as standard box sizes, size sorting of fruit, and grading of fruit for defects.

Some distributors are developing locally grown programs and indicating the availability of product by growing region. They may or may not pay a small premium for products from specific growing regions.

Prices that distributors pay to growers are often less than one-third of the retail price. This may sound like a big discount, but you have to consider the marketing and shipping services that these firms provide. Compared to selling direct, the grower has significantly lower labor and transportation costs. As a grower, you can move much more volume through a distributor, though at a lower price than if you were to sell direct. Even with this, your profits can still be higher due to the potential costs you avoid and the higher sales volume. Since distributors want product that is less ripe, many successful small producers sort out their harvest according to ripeness, sending their least mature produce to a distributor, packing somewhat riper produce in CSA boxes, and selling the ripest produce at farmers markets.

MARKETING COLLABORATIONS

As a small producer, you may find it beneficial to collaborate with other producers in some marketing-related activities. It may be as simple as arranging to trade off with neighboring producers on delivering your produce to your restaurant and distributor customers. Alternatively, a group of growers may form a cooperative or a limited liability company (LLC) to create a firm to pack, market, and deliver their products while still keeping their separate identities intact, or they may choose to consolidate their crops. The Community Alliance with Family Farmers (CAFF) has created regional grower's collaboratives by buying produce from numerous smaller family-owned farms and marketing it to institutional customers, while retaining each grower's identity during the process.

Traditionally, growers have formed cooperatives to assist them in marketing their crops. See the sidebar of "Tuscarora Organic Growers," to learn how collaboration can increase your market power and help improve your profitability.

REGULATIONS

It is essential that you educate yourself on sound food safety and postharvest handling practices for your products, as well as the related regulations. This information will affect your marketing practices, so make sure to read chapter 10, Postharvest Handling and Safety of Perishable Crops, before you finalize your marketing program. As a marketer, you also need to comply with regulations related to land use, labeling, and other factors, enforced by various levels of government. The following checklist will help you understand about some of these regulations.

- If you are marketing through a certified farmers market, you will need to get certification from your county agricultural commissioner.

- Check with your city or county's planning department about zoning laws that may affect where and how you can market your products, including marketing through farm stands.

- Depending on your products, you may need special permits from your county's health department or environmental health department and perhaps even from the California Department of Public Health.

- Assuming that your business operates under a fictitious name, you will need to make a fictitious business name filing with your county recorder's office.

- If you label your products with your business's fictitious name or some other fictitious name, you should also consider filing a trademark application

TUSCARORA ORGANIC GROWERS

Tuscarora Organic Growers, a Pennsylvania-based produce marketing cooperative, was founded in 1988. Twenty-eight farmer members and 17 nonmembers delivered approximately 100,000 cases of products—organic vegetables, flowers, eggs, and plants—to the cooperative in 2008. The products are marketed year-round under the Tuscarora brand—with the individual grower's identity preserved—in the Baltimore-Washington DC metro area to retail grocery stores, food co-ops, restaurants, and distributors. Many of the farmer members also sell their produce directly at farmers markets and through community-supported agriculture (CSA) programs.

In addition to its marketing function, the cooperative serves an important role in coordinating its members' production. Specifically, it advises members on what crops to plant and when to plan them in order to extend the marketing season and provide a steady flow of product into the marketplace.

The cooperative is democratically governed by its members; all of the board members are also producer members of the cooperative. Profits are allocated to members in proportion to their sales revenues.

By collaborating with one another through the cooperative, the growers gain market power; they are coordinating their marketing rather than competing against each other. They also enhance their attractiveness to customers by offering greater product variety, higher volumes and better customer service as a group. What's more, they gain economies of scale (and save their valuable time) by using the cooperative's hired staff to manage their Web site and handle their sales, deliveries, and billings. For more information, visit the Tuscara Organic Growers Web site at http://www.tog.coop.

with the California secretary of state's office (http://www.sos.ca.gov/business/ts). If you are also selling your products in other states, you should also consider a similar filing with the federal Patent and Trademark Office (http://www.uspto.gov/main/trademarks.htm).

- You need a produce or processor dealers license from the California Department of Food and Agriculture (http://www.cdfa.ca.gov/mkt/meb/index.html) if you buy produce or get it on consignment from other producers or if you process it for resale. You will also need a Perishable Agricultural Commodities Act (PACA) license from USDA (http://www.ams.usda.gov/AMSv1.0/fv) if you buy and sell produce from other growers.
- You will need a seller's permit from the California Board of Equalization (http://www.boe.ca.gov/info/reg.htm#sales) if you sell products that are subject to sales tax.
- Commercial customers who buy your produce can request that shipping point inspections of your produce be done in the field, at packinghouses, or at cold storage facilities (http://www.cdfa.ca.gov/is/i_&_c/spi.html). Inspectors can certify California-produced commodities as to quality (grade), condition, size, net weight, temperature, and other factors.
- Learn your rights regarding delinquent accounts and other unfair trading practices you may encounter when you sell produce. Check out USDA PACA Web site (http://www.ams.usda.gov/fv/paca.htm). If you plan to sell your produce to a new commercial customer, you can call PACA's toll-free number to review their PACA complaint and license history.
- If you are selling produce to wholesale produce buyers such as distributors and grocers, there are two online credit rating services that you might consider subscribing to: the Produce Blue Book (http://www.producebluebook.com/) and Red Book Credit Services (RBCS) (http://www.producecredit.com). Both specialize in providing credit information on firms (especially buyers) active in the wholesale produce industry. However, it is not likely that either service will have information on independent grocery stores, unless they operate at multiple sites.

- If you plan to market eggs, dairy products, poultry, fish, or meat products, you will have to meet additional state and federal regulations. You can start reviewing them at the appropriate section of the California Department of Food and Agriculture's (CDFA) Web site: http://www.cdfa.ca.gov/regulations.html. A University of California publication, *Selling Meat and Meat Products* (http://anrcatalog.ucdavis.edu/BeefCattle/8146.aspx), provides a summary of the regulations for meat.
- If you are marketing products labeled as organic, you must comply with the provisions of the USDA National Organic Program (http://www.ams.usda.gov/AMSv1.0/nop) and the CDFA Organic Program (http://www.cdfa.ca.gov/is/i_&_c/organic.html). Your farm and facilities will need to be certified as organic by an accredited certifying agent.

Last of all, remember to discuss your insurance coverage with your insurance agent as you expand your marketing efforts. In particular, make sure that you have adequate product liability coverage to protect your farming operation.

PROCESSED PRODUCTS

You can increase your revenues by processing your products in a variety of ways—such as cooking, pickling, juicing, drying, and smoking. Processing can open new markets for you, add value to your seconds, bring recognition to your farm, and extend your marketing season, perhaps even to year-round. But processing adds costs and requires more labor, management, and investment in equipment.

You also need to comply with numerous regulatory requirements. In California, food made or stored in the home cannot be sold to consumers. However, if you only handle in-state sales directly to the final consumer (such as at your roadside stand, CSA program, or farmers market), you are considered a foodservice or retail operation and can prepare food in a certified commercial kitchen. Exceptions to this are meats or meat-containing products or nonacidic or acidified canned products (such as green beans or pickled vegetables; see http://ucfoodsafety.ucdavis.edu/Food_Processing/Acidified_Foods/). Many farmers find it easier to rent an approved food processing facility such as a restaurant or a kitchen at a meeting hall rather than spend the $100,000 or more that it would take to build and maintain their own on-farm commercial kitchen. Some farmers contract with food manufacturers to process their products. Such businesses are called co-packers.

If you go beyond direct marketing for your processed products and sell them on consignment or through distributors and retailers, you will need to register with the California Department of Public Health Food and Drug Branch as a processed food manufacturer, packager, holder (http://www.cdph.ca.gov/programs/Pages/FDB%20ProcessedFoods.aspx). After you file your application with the Food and Drug Branch, they will do an inspection of the facility. This must occur before you can sell your product to other retailers. The Food and Drug Branch will review your business and product information, production and process controls, sanitation control procedures, and your product labeling and advertising. If deficiencies are identified, you will be charged for reinspection at the rate of $100 per hour. Currently, annual registration fees start at $348 for a facility with up to two employees (including the owners).

It is particularly important that you develop a business plan (see chapter 6, Farm and Financial Management) before you launch into processing your products. Your business plan should include all of your products, not just the processed ones. Carefully consider your goals for your processed food business: do you want it to become your primary source of income or a way to enhance profitability by using more of your harvested production? The processed food business is very competitive, and differentiation is critical. Check out similar products at grocery stores and online. Consider carefully whether the target market for your processed products will be the same as for your unprocessed product; if not, you will probably need to change your strategy and tactics. Review the U.S. Food and Drug Administration (FDA) food labeling regulations thoroughly (see http://www.cfsan.fda.gov/~dms/flg-toc.html). Project your cash flow using several sets of assumptions. Evaluate how much time you personally will have to dedicate to your new venture. Will you need someone to fill in for you in your regular production and marketing activities?

You should consider starting small, perhaps selling your processed products at farmers markets. That way you will have a track record when you eventually make your sales pitch to local grocers. Stephen Hall's 2008 book *Sell Your Specialty Food: Market, Distribute, and Profit from Your Kitchen Creation* addresses many processing, marketing, financing, and planning issues that small processed food marketers face. For example, the book includes a section on handling buyer objections that you may want to review before you make a sales presentation to a retailer. Don't forget to promote your product on your Web site and send out press releases as you develop new products.

MARKET PLANNING

Once you have digested the four P's and market research, you are ready to work on your marketing plan. It will be a more detailed version of the marketing section from your business plan. Your marketing plan is your guide to how you will structure your marketing mix. Its purpose is to give you a road map for your marketing efforts in order to enhance their effectiveness and save you time and money. With a thoughtful marketing plan, you can immediately disregard certain opportunities because you will recognize that they are not consistent with your marketing mix. You can start by following this outline for creating your own marketing plan:

- Identify trends in your product category
- Identify your target market:
    - Demographics and other characteristics
    - Industry or societal trends that affect your customers
    - Size of target market and growth projections
- Devise your marketing strategy:
    - Overall objectives and mission statement
    - Positioning relative to competitors
    - Marketing mix, including specific marketing programs (a marketing program consists of a set of coordinated activities designed to achieve a specific goal for a product or group of products)
    - Products
    - Pricing
    - Place (distribution channels)
    - Promotions, advertising, and other activities that inform, persuade, or remind potential buyers about your products
- Financial analysis:
    - Sales growth projections
    - Projected profit and loss for each product
    - Break-even analysis
    - What-if scenarios (sensitivity analysis)
- Evaluation:
    - Measurable goals
    - Success measures

Alternatively, you can purchase a software package to help you create a marketing plan, or you can follow the Small Business Administration's outline, available online (http://www.sba.gov/smallbusinessplanner/manage/marketandprice/serv_whatdoyouwant.html). Whichever method you use to create your marketing plan, you will want to cover the key areas discussed earlier in this chapter.

RECORD KEEPING

Though it can be painful, record keeping will pay off for you. Keep track of your expenses, including your marketing costs, and note what product or market channel they are related to. Record how much time you spend making CSA deliveries and driving to farmers markets, and how much product you donate. Measure how much of your crop you don't sell because of spoilage. Don't forget to include overhead expenses, such as insurance, taxes, and memberships. Keep track of your revenues, too, making sure to record prices received according to market channel and crop.

You work hard to get your customers. Keep a spreadsheet with their contact information, sorted by market channel. Make sure that you have complete information for your commercial customers, including their mailing and physical addresses. Ask your consumer customers for their e-mail addresses; they are great to have when you want to send out notices about new products, special events, and promotions. They will also come in handy if you decide to start a CSA or another form of direct marketing.

EVALUATION

One of the things that usually ends up at the bottom of a producer's list of things to do is evaluating the operation and planning for the future. You owe it to yourself to do a thorough evaluation of your efforts each year, to determine what you can add, drop, or tweak so that you can increase your profits during the next year while not working as hard as you did the previous year! Here are some suggestions for your evaluation:

- To compare the profitability of your products, allocate your expenses and sales revenues by product.
- To determine the profitability of the channels that you are marketing through, allocate your expenses and sales revenues by market channel.
- Review the marketing programs you proposed in your marketing plan and assess which ones worked the best for you.
- Review your customers' comments—both good and bad—and decide what adjustments you need to make to your crop selection, production practices, handling, labeling, packaging, and sales practices.
- Evaluate your customer base and determine whether you are reaching your target market effectively. What, if any, adjustments should you make?
- Do a quick audit of the marketplace and identify new competitors, distributors, and retailers that you need to pay attention to.
- Do a quick environmental scan and identify political, cultural, and technological forces that are likely to create opportunities or threats for you.
- Incorporate your "Aha's" from this evaluation into a quick revision of your marketing plan.

RESOURCES AND REFERENCES

Web Sites

Agricultural Marketing Resource Center (AgMRC). http://www.agmrc.org.

ATTRA – National Sustainable Agricultural Information Service. http://attra.ncat.org.

California Federation of Certified Farmers Markets (CFM). http://www.cafarmersmarkets.com.

Community Alliance with Family Farmers. http://www.caff.org.

LocalHarvest. http://www.localharvest.org.

North American Farmers' Direct Marketing Association (NAFDMA). http://www.nafdma.com/Public/Welcome.

Service Core of Retired Executives (SCORE). http://www.score.org.

Small Business Development Centers. http://www.sba.gov/aboutsba/sbaprograms/sbdc/index.html.

USDA Agricultural Marketing Service. http://www.ams.usda.gov/AMSv1.0.

USDA Rural Development, Value-Added Producer Grants (VAPG). http://www.rurdev.usda.gov/rbs/coops/vadg.htm.

USDA Western Region Sustainable Agriculture Research and Education (SARE). http://wsare.usu.edu.

U.S. Small Business Administration. http://www.sba.gov.

University of California Cooperative Extension County Offices. http://ucanr.org/County_Offices.

University of California Small Farm Program. http://www.sfp.ucdavis.edu.

Publications

Bachman, J. 2004. Selling to restaurants. ATTRA National Sustainable Agriculture Information Service. Pp. 16. http://www.attra.org/attra-pub/sellingtorestaurants.html.

Born, H., and J. Bachmann. 2006. Adding value to farm products. ATTRA National Sustainable Agriculture Information Service. Pp. 12. http://www.attra.org/attra-pub/summaries/valueovr.html.

Feenstra, G., J. Ohmart, and D. Chaney. 2003. Selling directly to restaurants and retailers. University of California, Sustainable Agriculture Research and Education Program, Davis. Pp. 5. http://www.sarep.ucdavis.edu/cdpp/selldirect.pdf.

George, H., and E. Rilla. 2005. Agritourism and nature tourism in California. University of California Division of Agriculture and Natural Resources, Oakland. Publication 3484. Pp. 159. http://anrcatalog.ucdavis.edu/SmallFarms/3484.aspx.

Hall, S. 2008. Sell your specialty food: Market, distribute, and profit from your kitchen creation. Kaplan Publishing, New York. Pp. 320.

Jolly, D., and C. Lewis. 2005. Food safety at farmers markets and agritourism venues. University of California Small Farm Center, Davis. Pp. 40. http://sfp.ucdavis.edu/docs/publications.asp?view=11.

Lobo, R., L. Lev, and S. Nakamoto. 2009. A market-driven enterprise screening guide. Small Farm Program Preliminary Publication. University of California, Small Farm Center, Davis, CA. Pp. 16. http://sfp.ucdavis.edu/docs/new_enterprise.pdf.

Rangarajan, A., E. A. Bihn, R. B. Gravani, D. L. Scott, and M. P. Pritts. 2000. Food safety begins on the farm: A grower's guide. Cornell University, Cornell Good Agricultural Practices Program, Ithaca, NY. Pp. 28. http://www.sfp.ucdavis.edu/pubs/articles/foodsafetybeginsonthefarm.pdf.

USDA Economic Research Service. 2009. Price spreads from farm to consumer: At-home foods by commodity group. http://www.ers.usda.gov/Data/FarmToConsumer/pricespreads.htm.

U.S. FDA Center for Food Safety and Applied Nutrition. 2008. A food labeling guide. September 1994; revised April, 2008. http://www.fda.gov/Food/GuidanceComplianceRegulatoryInformation/GuidanceDocuments/FoodLabelingNutrition/FoodLabelingGuide.

8
Labor Management

HOWARD ROSENBERG

Farm businesses differ with respect to their goals, strategies, production methods, use of resources, and economic results. All of them, however, small as well as larger, have in common a need for human work. Without the abilities and efforts of people, agriculture would not yield food or fiber. Nearly all farm operators depend on others to make their systems run, and most have to hire at least some labor from outside the family. But many are hard pressed to meet the challenges of employee recruitment and selection, compensation, supervision, and other day-to-day communication, all within a complex legal environment. Taking the time to consider these topics is an investment in smoother operations that is bound to yield better business and personal results. This chapter presents some guidelines that will help you make the countless decisions involved in managing human resources on the farm.

MANAGEMENT FUNCTIONS AND THE LEGAL CONTEXT

Success for a small farm takes more than the knowledge and skills that farmers long have applied to growing and selling agricultural products. Hard work and smart decisions about biological and marketing processes do not by themselves ensure good business outcomes. Another set of decisions, the ones that influence the people who perform work that is integral to the farm, is very consequential.

Returns from labor expenditures, including the unpaid hours that family and friends may put in, are hardly fixed by any formula. Two similar farms with identical payrolls and other inputs can have quite different results depending on how well the people at each farm do their jobs and work together, which in turn depends on how they are managed. Farm operators shape the contributions of others through relatively long-term planning and organizing decisions as well as more frequent staffing choices and day-to-day leadership and control practices. Regardless of how often decisions of all these types are made, their effects touch workers every day.

A farmer's labor management choices are constrained by a large body of federal and state laws that is formidable in its variety, intricacy, and propensity for change. Mostly designed to protect workers by controlling the actions of employers, agricultural employment laws carry requirements, restrictions, and rights. They set bounds on (1) standards for particular terms or conditions of employment—such as wage rates, rest periods, work hours for minors, and safety measures—and (2) certain interactions between employer and employee—such as processes of hiring, establishing contractual agreements, responding to complaints, and firing. Still other laws and institutions deal with the overall labor supply or workforce development outside of any particular employment context (e.g., public training and job search programs, health services, immigration policy).

The rules embody various definitions and coverage, and they are administered by several agencies with various levels of enforcement capability and orientations to agriculture. Some aspects of the employment relationships are covered by federal law, some only by California law, and some by both—in which case the higher standard, usually the state's, applies.

Agricultural employers, like all others, are obliged to report regularly to government agencies about their payrolls and employees and to meet various requests for other information. They have to withhold, report, remit, and maintain records of mandatory payroll taxes—income tax, Social Security contributions, and unemployment insurance tax. Partial exceptions apply to firms with very small yearly payrolls and to the earnings of an employer's minor children. Details on federal requirements, current rates, forms, and helpful guidance for farm employers can be found in Internal Revenue Service Publication 225, *Farmer's Tax Guide* (http://www.irs.gov/publications/p225).

State and federal agencies offer many other pamphlets and forms, printed as well as electronic, to help orient new employers and help all to meet their legal responsibilities. Virtually all now are accessible through Web sites. (See "Resources and References" at the end of this chapter for relevant links.)

BASIC STRUCTURAL CHOICES

The commodities produced, technology used, and scale of operation on a farm dictate roughly how much and what types of human work it needs. They thus affect workforce size, skills, stability, and supervision. The relationship between an owner and a production-level worker on a small family dairy with one year-round employee, for example, is much different than it would be in a multilocation tree fruit business that hires many people for a short summer harvest.

The workforce on many small farms consists of an owner, a few family members, and one or two hired employees. Duties there may be shared and job boundaries flexible, with the owner performing physical as well as managerial tasks. Even in such a business, where relationships are informal and fluid, benefits accrue from having a generally shared sense of their structure—the relatively fixed roles, relationships among them, and rules or norms that facilitate the division and coordination of work.

The definition of at least some structure in any organization can help prevent misunderstandings, personal frustrations, and other related problems. It is uncomfortable to most people as well as operationally inefficient if everyone in an organization has to start from scratch every day deciding who will do what. Consider spelling out for every person on the farm at least a brief description of general expectations, some routine duties or responsibilities, and other basic terms of employment.

As a farm business grows, it can benefit from a more explicit mapping of relationships on an organizational chart (see figure 8.1).

Just answering a few questions can help clarify your farm's organizational structure and may reveal areas of ambiguity or misfit that interfere with its smooth operation. Which people normally work together? Are some people responsible for what others do? Who reports regularly to whom? What can people in each job decide on their own? Is there an overall boss who has to clear all big plans or commitments? What other internal communications and working relations are most important to the farm?

These questions are especially important in a family operation, where anybody with the owner's surname might claim to have, and be feared as having, instant authority over any nonfamily employee. Having to meet the conflicting requirements of multiple bosses is difficult—if not impossible—and such a situation often leads employees to leave for better-organized pastures.

Figure 8.1. A sample organization chart.

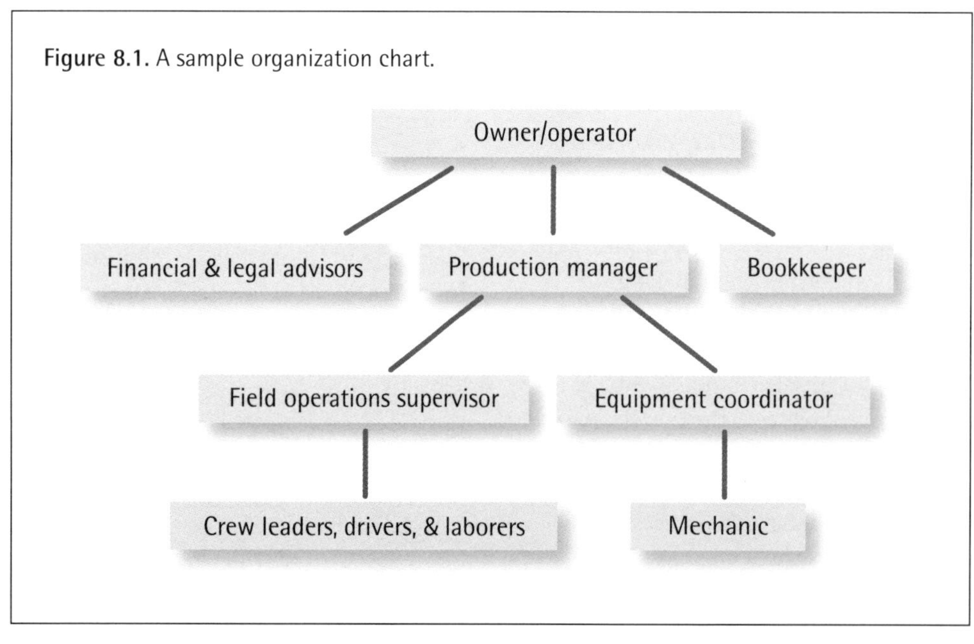

ENGAGING LABOR FROM CONTRACTORS

You can obtain nonfamily help by hiring people directly or by contracting with independent service providers, who themselves may employ others to do the work. While both forms of engagement are contracts, laws treat them differently.

When they buy services from a farm labor contractor (FLC) or custom operator, farmers can retain more organizational flexibility, more time for other, nonpersonnel management functions, and relief from the challenges of hiring, managing, and laying off the numerous employees needed for short-term seasonal work. As a kind of wholesale hiring action, engaging an FLC or custom operator removes many decisions from the farmer's hands, but it hardly eliminates all choices or all legal concerns. Just as the direct employment of a worker is subject to its own set of regulations, so too is the decision to do business with a contractor.

In many cases, the relevant enforcement agency will consider growers to be joint employers of FLC workers and thus jointly liable for any labor law violations—such as failure to pay the minimum wage, to carry workers' compensation insurance, or to withhold mandatory payroll deductions—that the contractor may commit. Generally, the more control a grower has over FLC employees' work process (as opposed to the work result) and the more economically dependent the workers are on the grower, the more likely their relationship is to be considered one of employment rather than independent contracting.

Farmers can reduce their risks by carefully selecting the FLCs they do business with, clarifying the terms of their agreement in advance, and sticking by those terms. There are documents to check and things to ask your FLC about before you shake hands on the deal or sign on the dotted line. Most critical are verification that the contractor has (1) a current California FLC license, (2) a current certificate of federal registration, with specific authorization for transportation and/or housing if she or he provides it, and (3) proof of workers' compensation coverage. Keeping a copy of the FLC's state license and verifying its validity with the state Division of Labor Standards Enforcement is now a requirement for hiring farms in California.

EMPLOYEE SELECTION

Why employ people at all? Farm owners give a variety of reasons:

- to use capital and overhead more fully
- to bring in expertise for a new technology or new crop
- to improve production quality or timeliness
- to reduce safety risks
- to support marketing and other managerial functions
- to free some time for family and leisure activities
- to reduce personal stresses on other staff

Whatever the impetus, the next issue is whom to hire, and it is critical.

No set of labor management decisions on a farm is more important than that by which the farm selects its employees. Each choice of a person to hire contributes to setting an organization's potential, and the processes of making that choice affects productivity, costs, morale, turnover, and the risks of various things going wrong. Like other decisions, employee selections can be made rather casually or through variously structured methods that may involve several steps. How much effort any particular selection process warrants will depend on the type of job to be filled and its expected duration. A decision about hiring a ranch supervisor, for example, calls for more time and attention than hiring a summer helper. In almost any case, however, it makes sense to start by clarifying the job to be filled, then to recruit applicants, assess how well they fit the job requirements, choose one (or more), and, finally, agree on terms of employment.

When hiring to refill an existing job, the initial step (clarifying the job to be filled) may be straightforward. But when the job is new or has previously been vaguely defined, the process of recruiting and selecting employees begins with several questions that boil down to: For what? What type of work is there to be done? How is it spread across various seasons? Where does the current staffing pattern fall short?

Once you have defined the farm's unmet needs, you can meet them through a range of possible arrangements, such as

- a short-term employee to perform specific projects such as yard cleanup, construction, and equipment repair
- a part-time, long-term employee to perform specific tasks on a regular basis, such as feeding animals in the morning
- a full-time, seasonal employee to help during the farm's period of peak production
- a year-round employee to perform various seasonal tasks throughout the year and supervise temporary employees during peak activity periods

Sometimes other adjustments can serve the purpose as well as or better than additional staff. If you change the assignments of people already in your employ, invest in equipment or training, or contract with an outside provider for specific services, you may find that option to be sufficient to rectify imbalances that have developed over time.

Fitting people with jobs involves working with good information about both. A job description is simply a verbal sketch of a given job: its purpose and content and the conditions under which it is performed. The description, particularly if it is in written form, lays a foundation not just for recruitment and selection but also for orientation and training, wage determination, performance review, and defensible termination, if necessary. It drives the collection of job-relevant information about potential hires and helps focus the hiring manager's communication with applicants.

The bundle of knowledge, skills, abilities, credentials, and other attributes needed to perform the job well is called a job specification, and information about whether a person fits it is best acquired through a combination of tools. The ability to follow written instructions, for example, might be assessed through a potential hire's completion of an application form; the knowledge and physical skills to correctly prune vines, through a practical test or demonstration in the field; the ability to calibrate chemical dilutions, through a math test; a willingness to work long and irregular hours, through an interview and reference check; an easy way with cows, through some observation in the milking parlor; and abstinence from drug abuse, through a medical exam.

Figure 8.2 presents a sample sequence of steps for gathering information about job applicants and acting on that information. It shows a preliminary interview, followed by completion of a written application form, employment test, in-depth interview, and reference check before the selection decision. A post-offer physical exam may be used to ascertain pre-existing medical conditions or to verify the prospective employee's fitness to perform certain tasks. In a small farm business, the prehire steps tend to be fewer, and the owner or general manager might handle them all. Although most farmers rely heavily on information and impressions drawn from an unstructured interview, the claims that applicants make about their abilities and experience often are not borne out by their performance on the job. For this reason, it is wise to plan on using at least one other tool to find out whether a prospect really has important job-related knowledge and abilities. Revealing job samples or practical tests can be devised for most agricultural jobs.

Pre-employment communication is mainly designed to obtain information for the employer's use, but applicants, too, need information and have to make decisions throughout the process. They also form attitudes that will affect whether they continue pursuing the job, accept it if offered, and perform reliably once in it.

Interviews and informal contacts during the selection process provide opportunities to both explain why an applicant would want to work for you and offer appropriate cautions. A realistic job preview that describes the more- and less-enjoyable aspects of the job will encourage many an unsuitable or unwary applicant to self-select out of the running. If the job requires long hours under the hot sun, endless repetitions of the same motion, fending off corny jokes from Cousin Louie, heavy lifting, tolerance of a sarcastic supervisor, strict punctuality, or concentration in the midst of noise, it is usually better to let the applicant know it right up front than to have your new employee suffer reality shock and leave you having to fill the job again.

A careful selection process naturally helps orient new employees to the job and workplace, but there are always things to learn and paperwork to complete beginning on day 1. The initial days are a time for forming impressions and attitudes that will color the employment relationship over the long term. Planning an introduction to farm facilities, co-workers, company policies, and job specifics is a courtesy that newcomers appreciate, and it accelerates their adjustment to performing at full capacity.

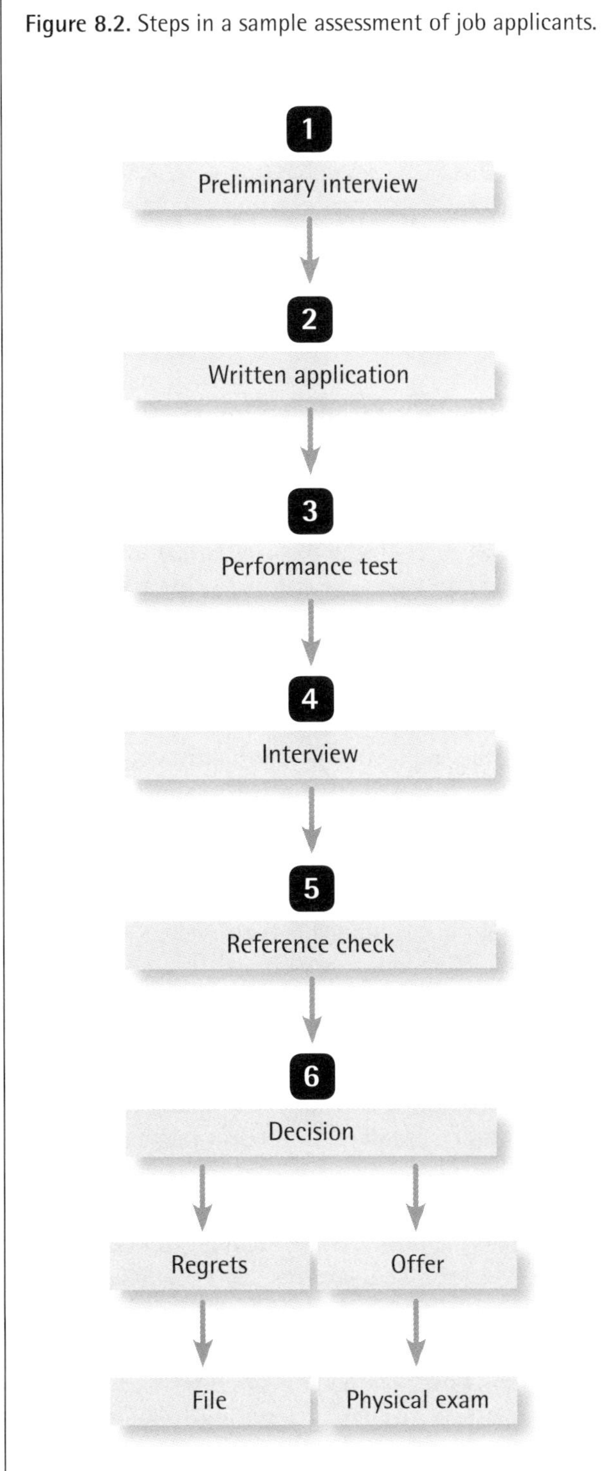

Figure 8.2. Steps in a sample assessment of job applicants.

Although family members who work on the farm may not be employees in a traditional or legal sense, managing to take good advantage of their capabilities involves largely the same processes as if they were. The needs to place people in roles that fit, to clarify duties and performance expectations, and to coordinate their work with that of others in the business are not limited to employees hired from outside the family.

COMPENSATION AND INCENTIVES

People generally work to obtain rewards that they value. To attract and retain good employees, a working environment has to offer rewards in reasonable balance with job demands and discomforts. Intrinsic rewards are those that people get from within themselves, such intangibles as satisfaction with a job well done, pride of workmanship, feelings of belonging, self-esteem, and a sense of contributing to a useful enterprise. Extrinsic rewards such as pay, fringe benefits, a new pickup truck, co-worker respect, a company vest, and control over certain farm resources come from the employer or others.

Despite the significance of intrinsic rewards, money is a special form of compensation for most employees—the main reason they work. It has value for both what it buys and what it represents as an index of personal worth. Other rewards count, but people tend to do what will get them more money. And to farm employers, wages and benefits paid are an important and significant cost of doing business.

Paying for performance. Just because money is a strong incentive to action for workers, that does not mean it always motivates them to do what the manager wants them to do. Managers may hope for one thing (e.g., many trees pruned) while their pay systems reward another (e.g., time on the job). But compensation systems can be carefully designed to align grower and worker interests by providing more dollars for desired performance in the short and long terms. The relationship between the sheer amount of wages paid and operational results is shaped by the choices managers make in structuring and administering pay, within legal bounds. Key decisions involve the following.

- **External equity.** What is the general level of pay on the farm, compared to others in the same labor market? A policy of keeping up with local rates is common, but some employers like to stay a cut above or below, and growth opportunities or other nonmonetary inducements can help make up for wages below the norm.

- **Internal equity.** What are the relationships between rates paid for work in different jobs within the company? Are mechanics, for example, paid the same as tractor drivers or irrigators? If not, how much more or less are they paid?

- **Pay basis.** On what basis is the periodic paycheck calculated for each job? Is a rate applied to units of time (e.g., hours or weeks) or to output (e.g., trays, cartons, bins, tons, sales, or vines)?

- **Individual wage determination.** Do all people in the same or similar jobs receive the exact same rate of pay, or do individual rates vary within a specified range? Can one hourly pruner, for example, make more than another? If a range of pay rates is used, what determines an individual's rate within it, and how and when may the rate be adjusted? Seniority and past performance can be systematically considered.

- **Benefits or indirect compensation.** What non-dollar or deferred pay (e.g., health insurance, living quarters, retirement income) does the employer offer to supplement current earnings, and who is eligible to receive it?

- **Communication.** How much and by what means are employees told about the overall pay structure? In some businesses, employees understand quite well how their pay is computed and what they can do to earn more. In others, the pay system is a mystery, whether because of inconsistencies within the system, a manager's belief that employees will work harder under uncertainty, or simply inattention to disclosure.

These choices affect both short- and long-run decisions by workers. In addition to revealing how jobs and individuals are currently valued, they inform employees as well as applicants about routes and limits of possible salary advancement. With a time-based wage structure, better performance may be recognized and translated into higher future earnings through two types of movement: (1) advancement in a pay range for a given job and (2) promotion to a higher-paying job. A clear system of pathways to job and wage growth, backed by fair administration, can powerfully influence workers interested in longer-term employment.

Incentive pay. Incentive plans aim to strengthen the linkage between performance and compensation by making the amount of current pay at least partly contingent on specific outcomes that the employer values. Piecework, the most common incentive pay plan in agriculture, has advantages and disadvantages. Multiplying a piece rate by the number of units produced has an appealing simplicity and equity on its face, as it clearly relates pay to actual work results. Piecework yields top dollar to top producers, and it gives employers advance certainty about direct labor costs per unit, so long as employees perform well enough to meet the statutory minimum wage rate. But output-based incentive pay may foster a rush for quantity at the expense of product quality, hurt crew cohesiveness in a few ways, upset traditional earnings differences between jobs, raise risks of worker injury from overexertion, and generate acrimony when rates are periodically set or adjusted.

Output-based incentive pay is best suited to situations where the output is easily measurable, employees have control of their production quantity, output quality is relatively unimportant or can be standardized through supervision, the employer needs to tightly control unit labor costs, and workers (or crews, if on a group piecework plan) work relatively independently from others. Pay incentives can be based on operational results other than sheer output—for example, gross farm receipts, net profits, equipment repair savings, average calving interval, crop yield per acre relative to a local standard, or milk production per cow. Keys are (1) what you value or want people to achieve, (2) what is under their control, and (3) what is measurable.

Like pay, fringe benefits can help a farm manager achieve multiple objectives: induce people to apply for the job, accept an employment offer, and stay in the job, produce high-quantity and high-quality results, come to work reliably, follow safety guidelines, cooperate with supervisors and co-workers, offer useful ideas, and not complain without good cause. Fringe benefits usually have more of an effect on employee attraction and retention, however, than on employee performance. Naturally, workers' relative preferences for different forms of compensation depend on their personal circumstances. Most single young men with healthy teeth who have a family in Mexico counting on them for income, for example, would rather have a higher hourly wage or a profit-based bonus than dental insurance and subsidized childcare.

SUPERVISION

Employees respond not only to incentives in the pay system but also to how they are treated from day to day. Most work under the personal influence of someone who provides supervision. From a worker's point of view, a first-line supervisor is his or her principal point of contact in the farm business. Through daily decisions and interactions, supervisors have an influence on whether capable people are attracted to the business, whether they stay as long as the employer wants them to stay, and how well they perform. Workers usually respond well to a supervisor who is clear about what they are expected to do and how, provides the tools and information needed to do it, shows them respect, and helps them develop as employees. Here are some typical employee descriptions of what a good supervisor does:

- provides me with a thorough introduction to the job
- teaches me specific job-related or technical skills
- makes sure I have specific goals in key performance areas
- assigns work that draws on my skills and strengths
- explains to me the formal and informal realities of advancement here
- makes sure I have opportunities to learn and develop my abilities
- gives me feedback and guides me on how to improve my performance
- offers opportunities to discuss operational problems
- asks for and listens to my opinion on pending decisions
- represents my interests and concerns to higher management

Farm owners and hired supervisors assign work to others to meet the farm's day-to-day operational needs. The very process of doing so takes time and effort, however, so entrusting employees with areas of responsibility or a regular set of duties rather than doling out tasks anew each day leaves farmers with more time to accomplish other things. Such delegation can be set up in a job description or a conversation, and it has the longer-term benefits of developing the workers' abilities and commitment. Delegation of responsibilities encourages employees to grow and apply their skills, serves as a step in training, helps prepare workers for more difficult jobs, increases the overall flexibility of the staff, and reduces the need to recruit more expensively from the outside.

Two common problems with delegation, however, are that (1) supervisors do not use it enough and (2) it may not work out well, which in turn feeds into the supervisors' reluctance to delegate in the future. One common reason supervisors do not delegate enough is that they have been burned in previous attempts. Here are some others:

- supervisors' uncertainty about their own role, authority, and extent of management responsibility
- need to feel control over results, and aversion to risk (i.e., "If you want it done right, do it yourself.")
- assessment of the supervised workers as unable to perform a job and handle responsibility for it
- fear that someone else's performance will be too good and show up the supervisor
- feeling that the owner wants the supervisor to stay physically active all the time
- greater familiarity, comfort, and self-assuredness in operational work
- desire to maintain an image and status of being overworked
- "Smokey the Bear" syndrome—satisfaction from stamping out forest fires on one's own (rewarding this kind of attitude may create an arsonist, metaphorically speaking)

What should a supervisor consider before delegation, and what may cause it to fail? Attributes of the worker, the supervisor herself or himself, and the nature of the work to be done all matter. If there is a tight deadline for completing the task, especially a complex task that challenges the worker's capacity, the chances of getting it done right are limited. A worker who misunderstands the assignment, lacks the needed skills, or has no desire to perform it will likely disappoint. And a supervisor can sabotage his or her own delegation effort by making a poor choice of employee to entrust with a task or by failing to carefully carry out the delegation process. Sometimes it really is better to do it yourself.

The information in table 8.1 suggests a method for deciding what to delegate. The table classifies bits of work according to their complexity (high or low) and frequency (one time or recurring). The tasks listed in area A are relatively easy and infrequent. Supervisors assign them as day-to-day needs arise. Tasks in area B—recurring and not complex—are usually delegated or simply identified as a regular part of the job. There is every reason to entrust these duties to nonmanagers for routine performance.

The complex, infrequent challenges of area C are the ones that managers are most inclined to either handle themselves or contract for someone with special skills to do. Sometimes these tasks can be structured into simpler component parts that can be assigned or shared by more people. Like those in area C, tasks in area D are complex, but because they come up often, there is greater advantage to having personnel on the farm who can be trusted to do them on a regular basis.

Delegation does not have to mean release of all control. Employees or family members can be entrusted with different degrees of responsibility for planning and performing work, as indicated by this range of approaches:

Degrees of delegation:

- Complete delegation
- "You handle the chore when the need arises."
- "Take care of it and let me know how it works out."
- "Work on it, and I'll check with you from time to time."
- "Let's talk about it and decide what to do; then you take over and report to me daily."
- "Give me your ideas. I'll consider them and then decide what we each will do."
- "I'll figure it out, do it myself, and explain it all to you later."
- None: "I'll take care of it. It's none of your concern."

A greater degree of delegation is generally more appropriate (1) for simpler, repetitive tasks and (2) when capable, reliable employees or family members are available to perform the work. As the capacity of the people on the farm's staff grows, duties can be delegated more completely and with greater confidence. One approach to fostering such development is to identify employees with greater potential and give them increasingly difficult and important tasks over time, perhaps using the delegation continuum above as a guide to gradually raising the bar. Most people respond well to being entrusted with responsibility that they feel capable of handling.

Table 8.1. Deciding what work to delegate

Complexity	Frequency			
	One time		Recurring	
Low complexity	A	Take that box to the house. Pick up sales agent at the airport. Tell Dad I'll be back late.	B	Check the tractor fluids. Milk the cows. Reorder when supplies dip.
High complexity	C	Create a safety program. Design a new field machine. Teach her how to drive.	D	Develop the ranch budget. Maintain the combine. Mix/apply the right herbicides.

PERFORMANCE MANAGEMENT AND REVIEW

Most people want to do their jobs well, and farm operators certainly need them to. Many workers, however, get few or mixed signals from managers about what they are supposed to do, how their work is seen, how they can do it better, and what the future may hold for them. Supervisors and employees alike find formal performance evaluation meetings stressful and try to avoid the loaded atmosphere typical of annual sessions in an office. But discussions about performance can be very useful and even enjoyable for both farmers and workers, especially when they occur in the flow of ongoing communication throughout the year.

The broad concept of performance management encompasses everything managers do to elicit good work from others. It includes, but is hardly limited to, assessing past work. Starting right from the identification of a job to be filled, growers may cultivate employee performance through their control of organizational structure, supervisory processes, and other work conditions. A systematic look back together at an employee's work after a period of time gives the manager and the worker information they need to diagnose performance problems and make adjustments.

At their best, performance review discussions are constructive exchanges that provide employees with more direction, guidance, and insight into their strengths and weaknesses on the job, and managers with ideas of how they can support better operations. If grounded in objective observations on job-relevant criteria, an appraisal can substantiate decisions on pay, promotions, disciplinary action, and other administrative moves, and reduce the risk of legal challenge if employees are discharged for cause. Without valid reviews of performance, both manager and employee are at a disadvantage when trying to learn from the past or plan for the future.

Regardless of whether they are conducted by an immediate supervisor and recorded yearly on a piece of paper, appraisals are informally rendered all the time. Assessments of performance may show in something as commonplace as a smile in the barn, a thumbs-up in the orchard, a scowl across the packing line, or good-natured needling in the greenhouse. Farm managers do not so much decide whether employee performance review takes place as they do what form it takes and how useful it is.

A farmer who opts for a relatively structured performance appraisal system faces three basic issues: (1) who does the appraising, (2) when the appraisal takes place, and (3) what the appraisal measures, and how. What constitutes a good choice on any of these three depends on what the manager intends to achieve with the appraisal. An appraisal intended purely to produce information for administrative decision making would focus on evaluation of past performance through rating and ranking procedures, with the appraiser acting as a judge and the appraised worker as a passive recipient. In contrast, an appraisal process designed as an aid to development and improved performance would focus on the future and would be more or less ongoing; the appraiser would act more as a coach or guide and the appraised as an active learner and co-planner.

When workers and supervisors see the formal review session as an ordeal to be endured, the performance appraisal process is likely to create hard feelings, erode relationships, and decrease morale. A tractor driver described his performance evaluation interview as "the worst day of the year. My manager sits there and tells me 'you didn't do this,' 'you didn't do that.' I leave feeling completely demoralized." Managers who want to prevent this kind of outcome and make appraisal meetings productive are advised to take these steps:

- Approach the session more as a coach, counselor, and planner than as an evaluator.
- Start off with nonevaluative sharing of information, inviting comments from the employee first. Emphasize the clarification of reciprocal expectations and future plans before moving on to assess the past.
- If you use a form, tailor it to the specific job, preferably one based on a written job description. Stationery store forms are too general and most are based on personal traits rather than specific job duties or results.
- Ask the employee to work with you in generating ideas for improving operations. Careful listening can be more productive than talking.
- Keep the performance review separate from any disciplinary action. Address any problem incidents when they occur rather than waiting until the appraisal meeting.

If the reviewer finds that the employee's performance has fallen short, effective corrective action starts with specifying how the employee's performance has been lacking and then beginning to figure out why. It may be easier to jump to the assumption that the worker is unfit or lazy, but you are more likely to uncover the cause(s) of poor performance and begin progress toward improvement if you consider such questions as these:

- Did anyone ever tell the employee what he is expected, or even allowed, to do?
- Does she understand enough of the big picture on a job to contribute as much as she can?
- Does he have the skill, physical energy, and time to handle his work load?
- In general, what difference does she think it will ultimately make to her if she does what is expected and refrains from what is prohibited?
- Do the company's policies or past practices indicate that good work improves job security? Current pay? Next year's pay? Future work assignments? Chances for layoff, rehire, or promotion?
- Does work performance affect whether immediate supervisors treat him with more or less respect?
- Do co-workers appreciate or resent her working well and according to standards set by management?
- Will he be expected (by either a supervisor or co-workers) to compensate for others' inadequacies?

Performance that fails to meet expectations can be sorted into three classes of cause: don't know, can't do, and won't do. The first is a matter of understanding. Managers have to communicate their expectations before the work begins if they are to reasonably expect the results they desire. Employees at all levels often cite insufficient direction as a major impediment to their performance and insufficient feedback as a block to the timely correction of problems. If workers are unsure about what is expected of them or do not know enough about the overall operation to do more than the minimum in their own jobs, the manager ought to find ways of delivering the missing information. Good opportunities to inform can be found in employee orientation, job descriptions, an employee handbook or written notices, rotation of job assignments, crew or staff meetings, and ongoing informal communications.

Once employees understand what is expected of them on the job, their performance depends on their ability and motivation—both what they can do and what they have a will to do. Neither of the two is sufficient by itself to ensure good work. Despite being extremely dedicated, enthusiastic, and hard working, a person with no mechanical skills will not be able to overhaul the tractor engine. But even a top-notch mechanic will not get the job done right unless he or she applies the necessary effort.

If employees do not have the knowledge or ability to perform effectively, a manager might consider restructuring the recruitment and selection processes for future hires. Possible adjustments range from posting more informative job ads in new places to utilizing practical pre-employment tests that help gauge actual current skills, initial deficiencies, and potential for on-the-job learning. Such changes have costs, of course, and these must be weighed, along with the costs of more fully training new hires or developing current employees.

Some farmers see prospective employees' character as more important than the set of work skills they bring to the job. Where the hiring decision is based heavily on such attributes as honesty, loyalty, integrity, responsibility, and learning potential, the employer has a greater responsibility for helping workers develop specific abilities once they are on the farm. Specific training after hire is needed everywhere to some extent, and it is a particularly good investment in productivity for the marginally qualified, for new workers whom you want to ease out of undesirable work habits or of particular techniques they may have learned elsewhere, and for people whose jobs on the farm have changed. Different tools or equipment can alleviate an employee's ability shortcomings by reducing the skill or stamina requirements of a task, and they may be considered reasonable accommodations that enable workers with disabilities to perform a job well.

A grower's choice of employee training techniques depends partly on the time and resources available, but most of all it depends on the training objectives—the type of skills or knowledge to be developed. Naturally, the best methods for teaching an employee how to use a computer, overhaul an engine, birth a calf, or identify voracious insects are not the same as those for how to lift the computer, change the tractor oil, pitch hay at the calves, or pick berries. The skills needed to perform more complex jobs or those that have more of a cognitive orientation necessitate more planning and time to help people learn.

Most skills training for physical farm tasks is provided informally on the job. Here is a good sequence of steps to follow:

- **Introduction:**
  - Explain the purpose of the task.
  - Find out what the learner already knows and can do.
  - Identify inputs to the task and expected outputs from it.
  - Relate personal experiences in learning and doing the job.
- **Presentation:**
  - Describe the procedure.
  - Demonstrate the correct performance of the procedure.
  - Point out critical decisions, tricky maneuvers, and foreseeable problems.
  - Invite questions.
- **Learner trial:**
  - Ask the learner to describe and perform the task (or parts of it).
  - Invite self-assessment and questions.
  - Confirm what was done well. Advise on what needs improvement.
  - Ask for a repeat performance.
- **Letting fly:**
  - Ask the learner if he or she is ready to perform alone.
  - Put the learner on his or her own, with the option to get help if needed.
  - Check back frequently to assess progress, coach, and ask and invite questions.
  - Taper-off coaching.

When workers know what to do and have the ability to perform but do not, the issue is rightfully viewed as a lack of effort or motivation. Some people call it an attitude problem. How much effort people exert depends to a large extent on what they expect to receive in return. Most employees work harder to obtain higher pay, greater job security, co-worker esteem, appreciation, or a kind word. They seldom put out additional effort in return for a blank stare, disrespect from the foreman, resentment from co-workers, and more work without compensation.

Farmers can use pay systems, performance appraisals, job allocation policies, and related supervisory practices to persuade workers that it really matters if they perform up to or above expectations. Analysis of what the employee stands to gain from working hard is a step toward uncovering conditions that lead people to work below capacity.

COMMUNICATION TO PREVENT AND SOLVE PROBLEMS

Communication with workers is integral to hiring and training them, helping them understand the operation and their place in it, assigning and coordinating their work, learning about their ideas and problems, building trust, and most other aspects of the working relationship between farmer and employees. Of course, communication among workers facilitates both performance and the social relationships that make the workplace more than simply where the job is.

Different types of information are needed at different times in the employment relationship. Some things are more important to discuss upon or before hiring (e.g., name of supervisor, normal work schedule, rate of pay), some at termination (e.g., reason for quitting or firing, last day of work, where to pick up a final check), some at regular intervals (e.g., which trees to prune today, hours worked by each crew member, how performance this year was evaluated), and some at irregular times (e.g., which cow showed signs of distress, where to meet the contractor, how to use a tool safely and what could happen if you don't). Some communication is legally mandated (e.g., deductions taken from gross pay, safe work guidelines).

Communication occurs through spoken word, written documents, and sometimes body language. Most managers use a combination of all of these. Many farmers have found written policies helpful to guide their own decisions and to inform workers what they can expect of the company. Policies can save administrative time, reduce employee uncertainties, foster consistency of management practices across supervisors and over time, and help avoid violation of a public law or personal sense of fairness.

Some businesses couple their policy statements with a code of conduct that describes standards of behavior at work—essentially, what is expected and what is prohibited. Having written rules and communicating them to employees both informs employees and supports disciplinary action, if necessary. Here are some topics that a code of conduct can address:

- Workday start, end, and break times
- Absenteeism, punctuality, and presence during work hours
- Use and misuse of equipment or supplies
- Endangering oneself or others
- Fighting and harassment
- Drugs, alcohol, smoking, and substance abuse
- Suggestions and complaints
- Misrepresentations (e.g., lying, falsifying documents)
- Personal activities during work hours
- Theft

With or without a code of conduct, any farm operator is bound to encounter worker performance shortcomings and misconduct. No matter how carefully a farmer designs jobs, selects employees, and trains, supervises, and pays them, things go wrong from time to time that call for a management response. Of course, there are different ways to deal with problem incidents that occur, and some of these not only affect subsequent behavior on the job but also exposure to legal challenges.

The possibility that they will need to defend against a charge of unjust dismissal is not the only reason for employers to think more than twice before firing. Growers usually do not want to get rid of people; in most cases they would rather live better with them. Discipline policies can help the grower do so by both clarifying standards in advance and structuring corrective action once problems occur. Applied consistently, this kind of policy provides a sense of fairness that colors the entire work climate. Good discipline means not always having to say "You're fired." In fact, it means rarely having to resort to dismissal.

Discipline policies vary with respect to length, precision, and content, but most have two major components: (1) a set of standards, rules, and prohibitions, and (2) a set of measures to be taken when the standards are violated. The list of standards communicates to employees what is and is not in accord with business needs, management's reasonable preferences, and the law. Consultation with other employers, an attorney, and current employees (especially supervisors) can help keep a manager from making rules that violate local norms, state or federal regulations, and workforce sensibilities. The second element is the action component, a set of guidelines for what happens when a rule appears to have been broken. If followed, these guidelines protect everyone from the risks associated with shooting from the hip; they also protect employees from arbitrary treatment. Prescribed responses typically include simple discussion—sometimes referred to as counseling—oral warning, written warning, suspension without pay, and discharge.

Progressiveness, another typical quality of discipline policies, is the application of increasingly severe measures in response to repeated offenses. For some offenses, such as violent assault, major theft, and deliberate damage to company property, the first incident is reasonable cause for immediate discharge. Most behaviors that violate standards of conduct or performance, however, do not warrant firing on the spot. First instances of tardiness, carelessness, indiscretion, or the like are usually considered cause for informal discussion or warning. More punitive responses, up to and including dismissal, follow repeated occurrences under a progressive discipline policy. Any offense, if repeated enough times, is sufficient to get a person fired. But at each stage prior to dismissal, the preferable effect is correction. Even when termination seems clearly warranted, a brief suspension before firing is often advisable to allow the employer to recheck all facts, assess the defensibility of the action if challenged, and consider broader business interests.

WORKPLACE SAFETY

Accidents at work hurt people and increase business costs. Employers are behooved, and also legally responsible, to have safeguards for workers on the job and in on-farm housing. Occupational Safety and Health Acts (OSHA) at both federal and state levels generally require that employers provide a safe and healthful workplace, and additional laws mandate hazard identification, employee training, and other specific measures to reduce risks of workplace injury and illness. California requires every employer of any size to establish, document, and implement an injury and illness prevention program (IIPP).

Field sanitation standards require that the employer provide toilets, drinking water, and hand-washing facilities for fieldworkers, plus training about the importance of good hygiene practices. Child labor laws give extra protection to minors by limiting both the type and hours of farmwork they may do. Detailed pesticide safety rules require training, field posting, and several other steps to protect workers.

In recent years, employees' rights to be free of sexual harassment have been bolstered under federal and state laws, and California employers are now obligated to provide biannual training to supervisors to help make the workplace safe in this respect. Also recently adopted and distinguishing California from other states are regulations that (1) limit hand-weeding or thinning in a non-upright position and (2) require specific measures to prevent heat illness in outdoor workplaces.

Despite these laws and all efforts to avoid mishaps, occupational injuries and illnesses do occur. Farm employers are required to help workers obtain medical attention when injured on the job and to record and report serious accidents. Agricultural employers in California are to carry workers' compensation insurance for workers, the same as in other industries.

Aches, pains, strains, and sprains—especially in the back, shoulders, hands, and arms—are probably the most common health problems for people who work on farms. Musculoskeletal disorders (MSD) can develop from the repetitive reaching, gripping, carrying, bending, squatting, kneeling, and twisting involved in many field and ranch jobs. They not only feel bad and reduce a worker's earnings, but also raise absenteeism, operational problems, and insurance premiums. A simple low- or no-cost change in task procedures, tools, or immediate environmental conditions can often reduce risks of MSDs by improving ergonomics, the fit between the job and a person's body and abilities (see U.S. DHHS 2001).

The federal Occupational Safety and Health Administration, its state counterpart, workers' compensation carriers, nonprofit educational organizations, and commercial service firms offer farm employers much help in meeting their obligations to provide a safe and healthful workplace. Technical references and ready-to-use training materials for farmers are available online. A consultation unit of the California Division of Occupational Safety and Health (http://www.dir.ca.gov/dosh/consultation.html) provides very helpful assistance at no charge, as well as publications (http://www.dir.ca.gov/dosh/puborder.asp).

RESOURCES AND REFERENCES

Agricultural Personnel Management Program Web site, University of California. Includes pages on California and federal laws, regulations, and agencies. http://apmp.berkeley.edu.

Billikopf, G. E. 2003. Labor management in agriculture: Cultivating personnel productivity. Second edition. University of California Division of Agriculture and Natural Resources and Agricultural Issues Center, Oakland. Publication 3417. Pp. 248. http://www.cnr.berkeley.edu/ucce50/ag-labor/7labor/001.htm.

California Chamber of Commerce. HR (Human Resources) California. http://www.calbizcentral.com/hrc/Pages/default.aspx.

Cal/OSHA. 2005. Guide to developing your workplace injury and illness prevention program. Cal/OSHA, California Department of Industrial Relations. http://www.dir.ca.gov/dosh/dosh_publications/iipp.html.

Hoppe, R. A., P. Korb, E. J. O'Donoghue, and D. E. Banker. 2007. Structure and finances of U.S. farms: Family farm report, 2007 edition. USDA Economic Research Service, Washington, DC. Pp. 58. http://www.ers.usda.gov/Publications/EIB24.

Pulakos, E. D. 2004. Performance management: A roadmap for developing, implementing and evaluating performance management systems. Society for Human Resource Management Foundation (SHRM), Alexandria, VA. Pp. 42. http://www.shrm.org/about/foundation/products/Pages/PerformanceManagement.aspx.

Rosenberg, H., et al. 2002. Ag help wanted: Guidelines for managing agricultural labor. Western Farm Management Extension Committee, Laramie, WY. Pp. 242. http://aghelpwanted.org.

Small Business Assistance Center, California Tax Service Center. http://www.taxes.ca.gov/Small_Business_Assistance_Center.

U.S. Department of Health and Human Services. 2001. Simple solutions: Ergonomics for farm workers. U.S. National Institute of Occupational Safety and Health (NIOSH). Publication 2001-111. Pp. 46. http://www.cdc.gov/niosh/docs/2001-111/pdfs/2001-111.pdf.

U.S. Internal Revenue Service. Farmer's tax guide. Accessible online. http://www.irs.gov/publications/p225.

Williams-Cosner, C., et al. 2005. Farm and ranch survival kit, Issue 3: Interpersonal relations. Oregon State University, Corvallis, OR. Pp. 9. http://extension.oregonstate.edu/wasco/small-farms-risk-management.

Small Farm Profiles

Mike Lee

A & M Farm

Manuel Bautista

Bautista-Gómez Farms

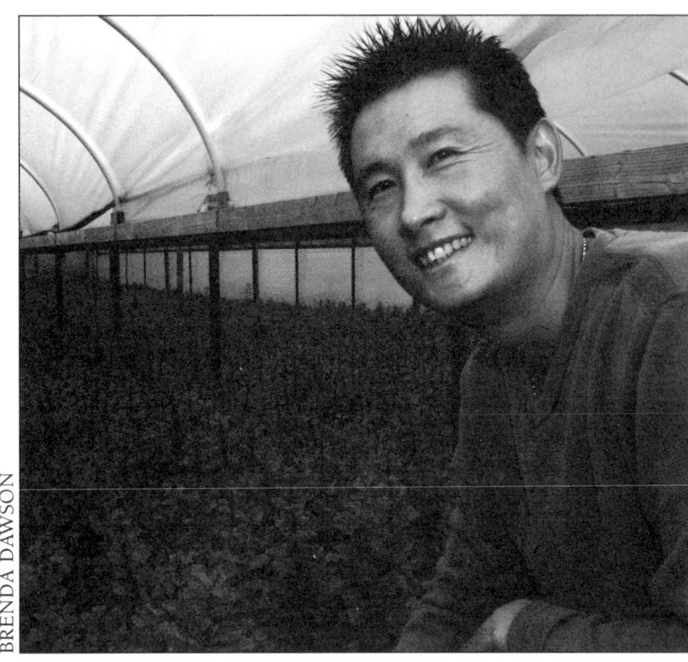
BRENDA DAWSON

SMALL FARM PROFILE:

Mike Lee

A & M Farm

"We grow Chinese vegetables that we sell to Chinese markets," explains Mike Lee, of A & M Farm in Morgan Hill. "My parents and the buyers understand each other, and that's why they started growing Chinese vegetables."

When he was 15, Lee moved to the United States to join his parents, who were already working on a California farm. The family decided to begin their own business, starting with greenhouse flowers. They soon switched to specialty vegetable varieties that are popular at ethnic markets—varieties that need the added heat and moisture a greenhouse can provide.

After attending college, Lee returned to help out and eventually manage the family farm, which operates on about 10 leased acres. Although he varies his vegetable crops each season, when we talked with him his primary crops were chives, ong choi (also called water spinach), yam leaves, snowpea shoots, and gao ge—a crop with which UC Cooperative Extension Farm Advisors were not familiar.

When choosing crops, Lee learned from his mistakes early on. For example, he once noticed prices for a vegetable were very high, so he jumped to plant 5 acres of it for the next season.

"But, unfortunately, everybody was thinking that was a good price," he says. "The whole area, everybody was growing that vegetable—and then couldn't sell any of it."

This particular vegetable, he learned, did not have a reliable market. Customers do not eat it regularly. Lee now has greater knowledge of both demand for different vegetables and annual market cycles.

With a mix of greenhouses, tunnels, and outdoor fields, Lee is able to time his harvests for early and late seasons when they can garner higher prices. For example, summer prices for ong choi are very near the break-even point, but earlier-season prices in April can bring a profit of up to 150 percent. When the price for ong choi falls in the summer, Lee sells other vegetables and then returns to ong choi in September, when prices are high again.

Lee also saves on labor costs by choosing crops that do not have to be replanted each season. His current crops have longer possible harvest periods so he can have them picked for specific orders.

"The vegetables that I'm growing can be harvested 1 week early or 2 weeks later, and won't be any problem," he says. "But when broccoli is ready, it flowers a week later, and then it's garbage."

Lee's knowledge of pricing and varieties comes from almost 15 years of managing his family farm's marketing. For years he made deliveries to stores in San Francisco's Chinatown every Monday, Wednesday, and Friday.

"My vegetables are not different from others," he says, "but you have to have the relationship with the buyers. I built up that relationship with the markets for years and years."

By making those deliveries, Lee developed another aspect to his farm business: sourcing vegetables from neighboring farms to sell to the markets.

"If I can find whatever the buyers need, every time they will call me first," he says.

Eventually he stopped making deliveries, and now he relies on his neighbors to take his vegetables to markets.

"It's definitely lowered profits," he says. "But that's the only way I can get away and plan other stuff."

In fact, Lee plans to return to China to expand his business. After 2 years of looking for land in China and setting up his new farm's infrastructure, Lee says the new farm will be about three times the size of his family's current farm.

"I am planning to go back to China because here the market is too small (for these crops)," he says. "It's not a good place to expand our farm."

—*Brenda Dawson*

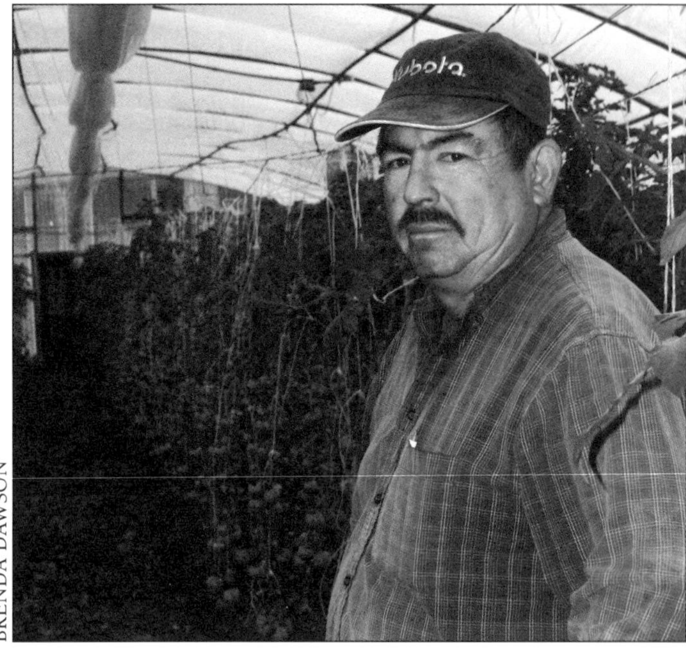
BRENDA DAWSON

SMALL FARM PROFILE:

Manuel Bautista

Bautista-Gómez Farms

"It doesn't feel quite right to say I'm living the American Dream," says Manuel Bautista, a farmer in Arroyo Grande. "Some people think, 'How come he has this much?' They're jealous, but I'm still working at 9 at night with my tractor lights on."

Bautista first came to California in 1983 and worked with his cousin, picking peas and riding the tractor. When his cousin quit, Bautista started farming on his own, growing squash and cherry tomatoes. A neighbor who spoke better English helped him get started at a farmers market in 1988.

"The first time my wife brought in a box of squash, she made $13 in two hours," he said. "We opened up to other farmers markets and started growing more things."

In 1993, Bautista started renting a tractor and the land he farms now. A few years later, Bautista bought the tractor. In 2000, when the owner decided to sell the land, he financed the sale to Bautista and his family.

Since then, Bautista has built two barns, a greenhouse, two tunnels, and a modular home at Bautista-Gómez Farms. He currently grows tomatoes, squash, salad greens, herbs, flowers, Oriental lilies, raspberries, blackberries, peas, and onions.

"Everything we have—cars, house, everything—has come out of this business. We don't have other income," Bautista says. "The best thing I've done is I've been investing in the farm every time I have extra money."

His products are sold primarily at 13 farmers markets every week, by his wife. They have seen the farmers markets change over decades of selling there. One trend Bautista caught on to early was growing salad mix, now sold by nearly every vendor at the market. Bautista charges about $1 more per bag than his competitors and takes pride in having the best salad mix, with a greater variety of leaves.

He cautions growers who want to enter a farmers market now, because there may not be open stalls or the market may already be flooded with common products.

"Before you spend one penny, you have to go to the manager of the market and ask," he says. "If not, you're going to be wasting money."

Bautista has learned to pay careful attention to his farmers market customers and what they want. He has also diversified his crops so he always has something to sell at the market, even if there's frost or bad weather.

"I waste money and a lot of time, but I'm trying. And I try and try and try, and sometimes, I'll get it right," he says.

Bautista sometimes has a few employees, but his own workdays during summer months are 14 to 16 hours each and were originally 7 days a week. After a decade of farming, the family vowed not to work Sundays anymore.

At 50 years old, Bautista started looking for ways to retire from farming at 60 or 65. He would be happy to see his children farm but thinks it is too much hard work. He chose to farm because he enjoys it, and his children will most likely find other jobs.

"I'm so proud," he says. "I paid for my daughter to attend college for 4 years, and she graduated. I gave her as much as I can. Now my second daughter is in college.

"I don't have $10,000 in the bank, but I've been doing the things that I'm not going to do if I work for someone else."

—Brenda Dawson

Small Farm Handbook

THE FARMING SIDE

THE FARMING SIDE

9
Growing Crops

BEN FABER, MARK GASKELL, AZIZ BAAMEUR, MARK BOLDA, STEVE TJOSVOLD, AND MARIA DE LA FUENTE

Growing crops on a small farm is often the most enjoyable part of a farm business operation, but marketing and market research are of equal or greater importance for the financial success and long-term survival of the farm. New and continuing small farmers quickly learn that they will spend as much or more time marketing their crops as they do producing them. Consistent and efficient production should not be neglected, however, and maintaining a close link between marketing and production is essential for small farm success. Many crops have fundamental physiological processes in common and share similar responses to soil, water, and climate conditions. In this chapter we provide some basic background information that is useful for a variety of crops and then go on to elaborate on some specific crop groups. Understanding these processes will help you plan your small farm crop operation and set priorities.

WATER AND IRRIGATION MANAGEMENT

Water Needs and Availability

Crop water needs vary with season, climate, soil type, plant type, growth stage, and rooting depth. However, a general estimate of crop water needs is the starting point for making critical decisions about the crop mix, planting and harvest times, and production areas to be planted on a farm. You need to plan an irrigation system with enough capacity to meet plant needs during the time of the year when water use is highest, even though that rate of water consumption may only be needed for a short period. Peak water needs must be matched by water availability.

In general, the absolute minimum continuous-flow per-acre water requirement from a well is 5 gallons per minute (7,200 gals per day) for drip irrigation, which will very likely be the most efficient irrigation method for your farm. If a house is served by the same well, the household will typically require an additional 2 to 3 gallons of water per minute. It can take about 7,000 gallons every day to irrigate an acre of mature fruit trees or vegetables in the heat of a Central Valley summer. In California, water availability and cost are important aspects of farming that must be factored into your economic decisions. Increasing water demands for urban uses and the possibility of water use conflicts between agricultural and urban communities may make low-volume irrigation the only realistic option for many new growers who are just getting started in farming.

To estimate your operation's water needs, determine your soil type (see Chapter 4, The Basics), consult your University of California Cooperative Extension Farm Advisor or the United States Department of Agriculture (USDA) Natural Resources Conservation Service (NRCS), and read about the water needs of your crops. Tables on the water needs of most crops in most regions of California are available from your UC Cooperative Extension Farm Advisor. Rooting depth and the evapotranspiration at a specific farm site are also important determinants of water needs. A crop's rooting depth determines how much moisture depletion it can tolerate before it becomes stressed. Evapotranspiration is the amount of water lost from soil and plants because of evaporation and transpiration. Evapotranspiration is usually reported in inches or millimeters per day or week.

Plant water requirements are given in acre-inches as well as gallons. In some foothill counties water is measured in miners inches (1 miners in = 11.22 gals per minute). The sidebar "How to Estimate Water Needs" will help you convert the figures to the measurements you use most often. An acre-inch is the number of gallons needed to cover 1 acre with water 1 inch deep, 27,154 gallons. You will need to do some math conversions to get from the evapotranspiration units of inches or millimeters to gallons, and eventually to gallons per day or per hour.

HOW TO ESTIMATE WATER NEEDS

To estimate water needs, select the peak water use period. For example, assume you will grow 5 acres of watermelons. Through research you have found that during the peak water use period this crop uses about 0.25 inch of water per day.

0.25 in/day x 5 ac = 1.25 ac-in of water per day per 5 ac

Now you need to match the amount of water you need to the well's hours of operation. If your irrigation system is designed to run 10 hours per day, start with this equation:

1.25 ac-in ÷ 10 hrs = 0.125 ac-in pumped per hr

One acre-inch is equal to 27,154 gallons, so your pumping need converts to 3,394 gallons per hour:

0.125 ac-in/hr x 27,154 gal per ac-in = 3,394 gal/hr

How many gallons is this per minute?

3,394 gal/hr ÷ 60 min = 57 gal per min (gpm)

But there's more. If the irrigation system has a distribution uniformity of 80 percent, you will need 20 percent more water in order to guarantee that the full water requirement is applied to every part of the acreage. In this case, you will need an actual flow of 71 gallons per minute.

57 gpm ÷ 0.8 (80%) =71 gpm

If your pump puts out 150 gpm, you can reduce the number of hours per day you run by about half, or if you want to skip a day, then you could run the full 10 hours and put on twice as much per irrigation. You can irrigate one-half of the acreage each day. This would allow you to use half as much water on any given day, or alternatively you can double the time of operation (if your system has been designed to do so). If the operating time is doubled, the required flow can be halved; in this example, to 35 gpm.

Water Needs and Climate

Total water requirements vary with climate as follows, depending on the crop and time of year:

- Plants in marine areas (fog-influenced climate where temperatures rarely exceed 75° to 80°F during the summer) use about 12 to 16 acre-inches of water per growing season.
- Plants in cool coastal areas (intermediate climate zone with some fog and a few hot summer days) use about 16 to 26 acre-inches of water per growing season.
- Plants in warm coastal areas (intermediate climate zone with little summer fog and several hot sunny days) use about 26 to 36 acre-inches of water per growing season.
- Plants in warm valley areas (interior valley climate with many summer days exceeding 90°F) use about 36 to 48 acre-inches of water per growing season.

Irrigation Systems and Management

Choose an irrigation system—surface, sprinkler, or low-volume (drip or microsprinkler)—that provides enough water and fits your crop management system. There is typically a trade-off between a system's initial cost and its recurring and maintenance needs. For example, drip and microsprinkler systems are more expensive to begin with, but they require less labor with each irrigation, do not need to be moved, and can be automated. It is important to compare the advantages and disadvantages of each system. In some cases, the water savings that can come with a low-volume system might be the deciding factor.

Surface irrigation. Any irrigation system that uses the soil surface to deliver water is called surface irrigation. Depending on the crop, region, topography, and water cost, you can use basin, border, or furrow irrigation, or a combination of these types. Surface irrigation is adapted to relatively flat ground and areas where large flow rates are available. Growers often use furrow irrigation for row crops, including vegetables and flowers. Basin and border irrigation are usually used for tree crops and forage or grain crops.

In basin irrigation, the water is ponded in small basins. There is little or no runoff, though on soils with a low infiltration rate the grower may

DAILY WATER NEEDS FOR VEGETABLES AND ORCHARDS IN SAMPLE CLIMATE AREAS

Climate conditions	Water needed per day per square foot*	Water needed for 1 acre of vegetables or mature orchard with a cover crop*
	- - - - - - - - - - - *gallons*	- - - - - - - - - - -
Early spring-late fall *Foggy, cool day;* *Evapotranspiration: 0.10 inch per day*	0.062	2,701
Spring or fall *Some fog, warm day;* *Evapotranspiration: 0.20 inch per day*	0.125	5,445
Midsummer *No fog, hot day;* *Evapotranspiration: 0.25 inch per day*	0.156	6,795
Midsummer *Very hot, 100°F, windy;* *Evapotranspiration: 0.30 inch per day*	0.187	8,146

*1 acre = 43,560 square feet.

deliberately drain water off after enough has percolated into the soil. Basin irrigation is simple and inexpensive.

In border irrigation, you build up parallel levees spaced 30 to 100 feet apart starting along the borders of the fields. The fields are graded with a slope to facilitate drainage. Water is supplied from a ditch or pipe along the upper end of the field and flows across the field to the lower end, where the runoff flows into a ditch. The grower keeps the water flowing until enough has percolated into the soil to meet irrigation needs.

Furrow irrigation is similar to border irrigation, except the water flows through ditches that are 6 to 12 inches wide and 3 to 12 inches deep.

Surface irrigation requires less investment in equipment than sprinkler or drip irrigation, but it can use more water if you do not recover and reuse the runoff. Tailwater from surface irrigation can be recycled if you capture the water from the downhill end of the field in a reservoir and either pump it back to the top end of the field for reapplication or channel it into a lower field.

Sprinkler irrigation. In sprinkler irrigation, water moves through pipes to the sprinkler head or nozzle, which ejects it into the air to fall to the ground at varying distances. Many growers use portable pipe sprinklers to germinate a crop before switching to a different irrigation form such as furrow or drip. With tree crops, some growers start the young trees with small basin irrigation and then switch to portable sprinklers for the larger trees. Some orchard growers only use permanent sprinklers.

Your choice of crops will help you determine whether to install permanent sprinklers, portable sprinklers, or sprinklers installed on short, movable hoses attached to permanent risers. An array of sprinkler types—including microsprinklers—is available to help you design an efficient system.

Permanent sprinkler and drip systems are convenient to use and often reduce labor requirements. Portable sprinklers can be an efficient option in a large orchard, since the grower would incur labor costs anyway from sending someone out regularly to check for sprinkler function, condition, and damage, even with permanent sprinklers. The cost to make these checks is a little less than the cost to move portable sprinklers, but that expense could be offset by the convenience of having the portable irrigation equipment available for other crops when it is not needed in the orchard.

The water pattern should cover the root zone and, to avoid runoff, the water should not be applied any faster than the soil can absorb it. This is particularly true on soils with a high clay content or on compacted soils. Besides runoff issues, excessive water may cause other problems such as root rot.

A sprinkler system is a good substitute for a surface system where the soil is shallow or highly variable in texture or where topography is irregular. Be aware, though, that sprinkler watering of some crops such as cucurbits and strawberries may increase disease problems. High levels of salts (measured as *EC* or *electrical conductivity*) in sprinkler water can damage leaves.

Energy use, water pressure, labor, and management vary with each system. All of these factors affect costs, and therefore affect the decisions a small-scale farmer has to make in choosing, installing, operating, and maintaining an irrigation system.

Low-volume irrigation (microsprinklers and drip irrigation). Microsprinklers and drip irrigation are two forms of low-volume irrigation. As a general definition, low-volume irrigation is any type of irrigation that delivers water at a slow, steady rate, near the crop plant. Low-volume systems often work where water pressure and volume limitations prevent the use of other systems. Both microsprinklers and drip systems are suitable for a wide range of slope, crop, and soil situations. The appropriate system should be designed for each specific case.

Microsprinklers are small, low-volume sprinklers designed to deliver water to a targeted area near the plant but to wet a wider zone than drip irrigation. For this reason, microsprinklers are most often used with orchard crops with their larger root systems.

Drip irrigation is most often used with vegetables and other row crops. Drip irrigation is the frequent, slow application of water to soil through emitters that are located at selected points along continuous tape or tubing lines in vegetables or berry crops. When drip is used in tree crops, multiple lines and multiple emitters are used for mature trees in order to wet a larger soil volume. The increased number of emitters requires that the grower devote more attention to clog prevention. You can install emitters in polyethylene tubing or purchase tubing or tape with emitters pre-installed at set spacings. Different types of emitters are available to suit distinct crop and soil needs. The amount of soil wetted by drip irrigation is much less than the amount wetted by other irrigation methods.

In row crop situations, the seeded or transplanted crop is often watered initially with sprinklers that are then replaced with drip after young seedlings are established and the surrounding soil is uniformly wetted. This sprinkler watering is valuable in that it thoroughly wets the volume of soil surrounding small root systems, removes air pockets, and assures good root-soil contact. For newly planted trees, only about 10 percent of the soil needs to be wetted. Observations show that at least 33 percent of the soil in the root zone under mature crops should be wetted and that crop performance improves as the amount wetted increases, up to 60 percent or more. The amount of soil wetted depends on soil characteristics, emitter flow rate, irrigation running time, and the number of emitters you use. Newly planted trees may need only one emitter each, while a larger tree could use eight or more. (Large trees are sometimes watered with two minisprinklers that disperse the water over a larger root zone area.) With a drip system, you have to make frequent applications of water to dilute salts and move them away from plants' root zones and to supply the plants with their daily water needs. The periodic injection of acid into the system to dissolve accumulated salts that can clog emitters is also part of microirrigation system maintenance.

Benefits of low-volume irrigation systems. Low-volume irrigation systems generally use less energy and require a lower water pressure than surface or movable sprinkler systems. Drip and microsprinkler irrigation systems allow greater control over the placement and timing of fertilizer applications and may lead to more efficient fertilization. Highly soluble fertilizers or finely filtered organic fertilizers can be injected directly into a drip irrigation system, avoiding the need for additional equipment and labor for application. It is important to note, however, that these systems require just as much management skill as other types of systems, if not more.

Drawbacks of low-volume irrigation systems. One big drawback to drip and microsprinkler irrigation systems is their higher initial cost compared to surface and movable sprinkler systems. The disposal of used and discarded drip tape also presents a potential drawback to drip irrigation, but irrigation hose manufacturers are increasingly offering complimentary recycling services. Additional problems include trouble with clogged emitters, poor uniformity of water delivery if the system is not designed well, and salt accumulation in the soil.

Algal growth in the drip line may cause clogs or more expensive damage to elaborate drip systems, but this sort of problem may be avoided through periodic injection of chlorine into the system. All of these problems are easily solved or avoided, however, with careful management. Other problems with drip systems include damage caused by animals and insects that chew drip tape and ruptures caused by mechanical weeding equipment, both of which necessitate repairs. Finally, because water runs to the lowest spot in a field, you have to make sure that the ground is reasonably level or take special care to design the system to compensate for uneven pressure or flow problems. Pressure-compensating emitters may be necessary for some systems if the terrain is too uneven.

Irrigation distribution uniformity. In all cases, it is important for irrigation application to be uniform. Irrigation distribution uniformity (DU) is a measure of how water is distributed in the field. A field's DU value is presented as a percentage. A perfect system would have a DU of 100 percent, but such a system could not exist in the real world. An excellent irrigation system would have a value in the 90 to 95 percent range. Values above 85 percent generally reflect very well managed irrigation systems.

Measuring distribution uniformity of pressurized systems. The way to measure distribution uniformity for a pressurized irrigation system is to select a specified number of emitters and measure their output. If you have 100 emitters, lay out an evenly spaced grid across the field so that all parts of the system can be sampled. Identify a minimum of 12 emitters to sample, and keep in mind that the more emitters you sample, the more accurate will be your measurement of DU. The method described below works best if the number of emitters chosen for sampling is a multiple of 4.

Turn the system on, and upon going to the first emitter you have chosen for sampling, invert the emitter over a graduated cylinder or some kind of measuring cup and capture the water for a specified period of time, such as 20 seconds (the period used in the example that follows). Make a note of the amount of water in the sample. Then empty the graduated cylinder or measuring cup and go on to the next sampling emitter. After you have collected and recorded samples from all of your selected emitters, arrange the amounts from low to high, add the values, and calculate the average. Next, average the amount that comes from the one-quarter of the emitters with the lowest output. Divide this result by the average output of all the emitters and express the answer as a percentage (e.g., 0.80 = 80%). If DU is below 80 percent, take appropriate action, such as resizing pipe and doing more to correct for poor pressures in the lines. If it is greater than 80 percent, you may still be able to improve the system's efficiency by making sure all of the emitters are of the same make and output and that the filter is flushed more frequently.

Calculation example. Martha measures 12 emitters and finds 8.5, 8, 7, 8, 7.5, 6.5, 7, 7.5, 8, 8, 8.5, and 7.5 fluid ounces in her graduated cylinder after 20 seconds of capture. The values are arranged from low to high, summed, and averaged as shown in table 9.1.

Using the average outputs calculated above, divide the low-quarter average by the all-emitters average and render the result as a percentage to get your distribution uniformity (DU). In this example, Martha has a system DU of 88 percent (6.8 fl oz ÷ 7.7 fl oz = 0.88 or 88%). That is not a bad number, but if the field is on flat ground,

Table 9.1. 20-second emitter samples from the calculation example in text, arranged from lowest to highest sample amount

Sampled emitter	All sampled emitters (fl oz)	Low quarter of sampled emitters (fl oz)
1	6.5	6.5
2	7.0	7.0
3	7.0	7.0
4	7.5	
5	7.5	
6	7.5	
7	8.0	
8	8.0	
9	8.0	
10	8.0	
11	8.5	
12	8.5	
Total sampled amount	92.0	20.5
Average output per emitter	7.7 fl oz (92.0 fl oz ÷ 12 emitters)	6.8 fl oz (20.5 fl oz ÷ 3 emitters)

there could be room for improvement, such as walking the lines more frequently to check for clogging, or flushing lines more frequently.

SOIL AND FERTILITY MANAGEMENT

Soil provides physical support for plants and acts as a reservoir for water, air, and nutrients. Study your soil's texture, structure, fertility, and especially the amount of time it takes to drain. Soils that do not drain well may need to be made up into raised beds for row crops or berms for orchard crops as a way to improve drainage. In deep, well-drained soils, trees are less likely to get root rot and are likely to produce more fruit or nuts. On uneven or hilly terrain, pay attention to the need to reduce the potential for erosion. If possible, plant on a contour and use cover crops to help reduce potential erosion. Soil compaction can also lead to erosion. Avoid or reduce compaction by cultivating soils when they are not too wet, usually 3 or 4 days after a rain or irrigation. Determining the ideal time to cultivate a soil requires experience, but it is very important for seedbed preparation and for avoiding tillage problems.

Soil Texture

A soil's texture is determined by the mixture of different sizes of minerals in the soil and can be broadly described as coarse (sand), medium (silt), or fine (clay). The finer the soil's texture, the higher its water retention (or water-holding capacity), and higher water retention means longer intervals between water applications. Soil that is high in clay (heavy-textured soil) is sticky, hard to dig or plow, and dense, and its surface will crack in hot, dry, sunny weather. Sand (light-textured) soil is relatively infertile, and water can move through it quickly, removing soluble nutrients. Silty soils have a slower water intake rate than sandy soils but their water-holding capacity is higher. Most agricultural soils are loams, combinations of sand, silt, and clay. Loams are easier to work, more porous, and less likely to crack than clay, and they retain moisture and plant nutrients better than sand.

WHY IS DISTRIBUTION UNIFORMITY IMPORTANT?

Distribution uniformity (DU) is critical to the management of irrigated crops. When an irrigation system has 95 percent DU, that means that the same quantity of water is delivered to 95 percent of the field. This means 5 percent of the plants are not getting the same amount of water as the rest of the plants. 2.5 percent of the plants are getting more than the average, and 2.5 percent are getting less than the average. In order to make sure that the 2.5 percent of the plants are getting an adequate amount of water, it means the whole system run time should increase 2.5 percent to cover the underirrgation. That means that the 95 percent that were getting the same amount of water are now going to get 2.5 percent more than they were getting on average. You can make up for the reduced application in the remaining areas of the field by applying about 5 percent extra water through the system.

Uniformity is also important when you are injecting nutrients and other chemicals into the irrigation water. A uniform system will distribute fertilizers and other additives equally to all plants in the field. Uneven DU of water and injected nutrients is the major cause of uneven crop growth as manifested by undulating plant growth heights along the rows.

Water Conversion Factors

Area:
1 acre = 43,560 square feet

Concentration:
One part per million (1 ppm) = 1/1,000,000 = 1 milligram/liter

Flow rate:
1 cubic foot/second = 449 gallons/minute or 26,940 gallons/hour
1 gallon/minute = 0.01 acre-inch/4.5 hours

Volume:
1 acre-foot of water = 12 inches over a 1-acre field = 325,851 gallons
1 acre-inch of water = 27,154 gallons
1 cubic foot of water = 7.48 gallons

Soil Structure

Soil is also affected by the amount of organic matter it contains, which affects tillage and cultivation practices. An ideal soil for plant growth will contain 25 percent air, 25 percent water, and 50 percent minerals and organic materials. How soil particles are arranged determines its structure. When particles are aggregated, there can be channels between the aggregates to help let water in and permit gas (oxygen) exchange, both of which are important for plant growth and vigor. Soil organic matter can act as a glue to help aggregate soil particles. A grower can increase soil organic matter by adding compost or incorporating a cover crop to improve soil tilth and overcome some of its texture-associated limitations (e.g., poor drainage, poor water retention, or poor nutrient retention).

Fertility

Plants require a number of elemental nutrients for optimal growth. They get carbon and oxygen from the air and the other nutrients from the soil surrounding their roots. A fertile soil provides the nutrients a plant needs. You can improve the soil's fertility by adding appropriate nutrients and amendments.

Plants need nitrogen, phosphorus, and potassium in relatively large quantities. They also need smaller amounts of sulfur, calcium, magnesium, iron, zinc, manganese, copper, chlorine, boron, and molybdenum.

Soils that have an appropriate amount of organic matter are more fertile and produce better crops than those that do not. A grower can incorporate organic matter into a soil by plowing-under cover crops or crop residues or by adding organic soil amendments and fertilizers. Manure and compost are commonly added to improve soil fertility, and a variety of other types of organic soil amendments are available that vary widely in cost and effectiveness. Rotating complementary crops or crops from different plant families in a field (e.g., legumes followed by tomatoes, followed by cucurbits) and incorporating their residues into the soil adds organic matter and may have additional benefits for managing soil pests, maintaining soil tilth, and reducing plant insect and disease problems.

Soil Nutrients

The key to good plant health is a balanced diet and regular feeding with an appropriate amount of nutrients. Excess nutrients can be detrimental to the crop and can pollute groundwater, streams, and marine environments. Plants use most nutrients in the ionic or elemental form (e.g., nitrate [$NO3^-$], ammonium [$NH4^+$], potassium [K^+], and phosphate [$H2PO4^-$]), not complex molecules such as proteins, carbohydrates, or vitamins.

Nitrogen. Nitrogen is absorbed by plants as nitrate and ammonium. Nitrate is very soluble, and ammonium is rapidly converted to nitrate in well-aerated soils above 50°F. Nitrogen is stored in soil organic matter (living, dead, and composted microorganisms and plants) or can move freely in the soil solution. Some nitrogen also sticks to soil particles, and if the soil dries out that nitrogen can be lost into the air as a gas. Organic matter is important in nitrogen management because it allows the nitrogen to be released slowly, providing a continual, small supply for plants and serving as a buffer against leaching. In the absence of large quantities of organic matter, water management and nitrogen management (including the method of application, timing, and amount) become even more critical.

Phosphorus. Since phosphorus is not very soluble in water, it tends not to move much from where it exists naturally in the soil, where it is added as fertilizer, or where it is incorporated as part of organic matter. Generally, plant roots must grow to the phosphorus rather than having phosphorus move to the roots, so it is best to apply phosphorus close to roots. Phosphorus deficiency is rare in tree crops because the trees' extensive root systems generally reach the element wherever it is. Phosphorus can be leached from the root zone into groundwater in very sandy soils or can move into surface waters attached to eroded soil particles. Since little of the phosphorus in a soil moves anywhere or is removed by most crops, it can build to very high levels in a field that is repeatedly fertilized with phosphorus for a number of years. Once enough phosphorus accumulates in the soil, further applications are not necessary.

Potassium. Many vegetables are heavy users of potassium; potassium is often important in keeping fruit firm and extending its shelf life. Many tree crops require potassium applications. Potassium is water soluble but adheres to soil particles, so it does not move with water once it gets into the soil. Except in very sandy soils, leaching is usually minimal. Most California soils are naturally moderate to high in potassium.

Soil Sampling and Testing

Have your soil tested by a reliable soil-testing laboratory to determine the amounts of various elements it contains and to determine its acidity (pH) and electrical conductivity (EC). Soil tests can reveal nutrient deficiencies and point out areas that are high in nutrients. Most tests show soil pH and the amounts of nitrate, ammonium, potassium, phosphate, and minor nutrients. Contact your UC Cooperative Extension office for a list of soil-testing laboratories in your area. Your UC Cooperative Extension Farm Advisor can also help you interpret the soil test results.

Since soil types can vary over a field, you may need to take several samples from each field, usually one sample (consisting of 20 cores or slices of soil) for every 15 to 20 acres. Areas with different soil types or different fertilizer applications or other amendments are sampled separately. Avoid smaller atypical patches in the field when gathering samples. Samples of the surface soil gathered by simply dumping a few handfuls of soil into a bag will not provide you with accurate or helpful results. Take your samples after disking the preceding crop but before you fertilize for your next crop, and after irrigation.

You will need a clean plastic bucket or bag and a soil probe if you have it. If you do not have a soil probe, use a spade or trowel. Scrape away any organic debris on the soil surface and use your soil probe, spade, or trowel to take a sample that goes 8 inches deep. If you are using a spade or trowel, take a ½-inch-thick slice 8 inches deep and then trim away the sides of the slice to leave a 1-inch-wide strip. Put the core or strips into the bucket or plastic bag, break up the clods, and thoroughly mix the soil. There will be about 1 quart of a well-mixed soil in each 20-core sample you send to the laboratory.

Label the sample with your name, address, and a sample number. Keep a record of the sample number, the origin of the sample, any past crops grown in the field, the quantity and types of fertilizer used on previous crops, herbicides used, and any nematode or disease problems you know of.

Make sure the laboratory you select to process your soil samples uses University of California test methods or protocols. Out-of-state laboratories can be used as long as they follow the same protocols.

Leaf or Tissue Sampling

In contrast to methods used for vegetables and other annual crops, leaf sampling is used to guide fertility management in tree crops or longer-season and perennial row crops such as caneberries or strawberries. Mature leaf blades are taken from trees of a similar age, variety, and rootstock growing on similar soils. The leaves are sent for analysis to a laboratory, where technicians analyze the samples for nutrients and prepare a report for the grower. For specific guidelines on sampling and sample preparation, contact the laboratory that will be doing your analysis.

Cover Crops

Some growers will plant a field to cover crops between plantings of marketable crops as a way to add organic matter and nitrogen to the soil, utilize excess nutrients, and protect soil from potential erosion. Planting cover crops is considered to be a type of conservation practice that, when implemented, can be especially important for water quality protection. In some circumstances (e.g., mustards planted as a cover crop), cover crops also offer an excellent level of weed suppression and some disease suppression to benefit the succeeding marketable crop. Cover crops have been used in California agriculture for more than 100 years. Examples of annually sown cover crops grown in California include vetches, fava beans, mustards, and cereals. Perennial cover crops, which include clovers and perennial grasses, do not require seeding on a yearly basis.

Benefits. Both legume and nonlegume cover crops take up soil nitrogen that then becomes available to the following crop when the cover crop is cut and incorporated into the soil. Legumes also fix atmospheric nitrogen, which can contribute additional nitrogen to the following crop. Cover crops can be grown summer or winter; however, most California growers use winter species. Cover crops offer a practical means of supplying organic matter to soil to improve soil physical and chemical properties. The decaying organic matter can provide nitrogen and other soil nutrients for succeeding crops. Cover crops also provide habitats for beneficial insects, improve soil tilth, facilitate water penetration, and increase the diversity of microflora in the subsoil.

Drawbacks. Cover crop planting may have some negative impacts as well. Cover crops are often grown in the off-season—between crops—but many growers in mild production areas depend upon early and late season

FACTORS INFLUENCING NUTRIENT NEEDS

Factor	Influence
Aeration	Roots require oxygen for respiration and nutrient uptake. Heavy clay soil may lack sufficient aeration for vigorous plant growth.
Light	Photosynthesis provides the energy for nutrient uptake. It cannot happen without light.
Microorganisms	Soil microbes are important to soil fertility because they break down organic matter into usable nutrient forms. Some form a symbiotic relationship with plants (e.g., mycorrhizae and rhizobia). Soil microbes may also compete with plants for nutrients during decomposition of organic matter.
Moisture	Roots do not grow toward water, they grow in moist soil. Nitrogen moves with water, whether toward or away from plant roots.
Nutrients	Increased nutrient concentrations in the soil solution or on the soil particle surface result in increased nutrient uptake.
Organic matter	Organic matter provides a reservoir for nutrients over a long period of time and buffers against leaching of nutrients.
Pests and diseases	Pests and diseases prevent plants from absorbing nutrients efficiently.
Root systems	Most vegetable crop root systems are shallow. Crops with extensive root systems are more efficient gatherers of nutrients.
Soil pH	Nutrient availability varies with the type of soil. Some nutrients (phosphorus, iron, manganese) are more available in soils that are more acidic. Others (calcium, magnesium, molybdenum, sulfur) are more available in soils that are more alkaline. Soils with a pH lower than 6.0 or higher than 8.0 may need additions of lime or sulfur, but some crops (e.g., blueberries) do require more acid soils.
Soil type	Soil types are classified as sand, silt, or clay. Clay soils have a greater capacity to store nutrients than sandy soils, but they may drain poorly and are more difficult to manage during preparation for planting.
Temperature	Your crop fertilization program should be based in part on climate and weather. Crops roots and tops grow more quickly at higher temperatures, increasing their nutrient needs. Roots absorb phosphorus poorly at low temperatures, so you have to apply more phosphorus fertilizer close to the roots in cool or cold soils.

NUTRIENT NEEDS

Nutrient	Heavy users	Moderate users	Low users
Nitrogen			
Annual crops	Bok choy Broccoli Cabbage Carrots Cauliflower Onions Potatoes Sweet corn Winter squash	Cantaloupe Cucumber Lettuce Peppers Summer squash Sweet potatoes Tomatoes Watermelon	Beans
Perennial crops	Almond Berries Citrus Nectarine Peach Walnut	Apple Apricot Cherry Chestnut Fig Pear Pistachio Plum Prune	Filbert Grape Kiwi Olives Persimmon Pomegranate Quince
Phosphorus			
Annual crops	Beets Broccoli Cabbage Carrots Cauliflower Celery Lettuce	Beans Cantaloupe Cucumber Onions Peppers Potatoes Summer squash Sweet corn Tomatoes Watermelon	
Perennial crops			All fruit trees
Potassium			
Annual crops	Beets Carrots Celery Potatoes Winter squash		
Perennial crops	Berries Grape Prune Walnut		

crop production since prices are more favorable then. Cover crops may attract pests and require additional seed purchase, irrigation water, and labor. Fall cover crop plantings prevent growers from working up the ground in advance of spring planting. Additional time may be required following a grass or cereal cover crop so the crop residue can break down sufficiently to allow for seedbed preparation. In some areas (e.g., semi-arid regions of the Central Valley) it may not be practical to use cover crops because they may require substantial amounts of water for irrigation. Also, revenue-producing acreage is reduced when you grow a cover crop. However, you can also think of the cost of planting and maintaining a legume cover crop as the cost of producing nitrogen and improving soil quality. Given the current and probable future fertilizer price increases, it makes sense to strike a balance between applied nitrogen and cover crop nitrogen.

For organic growers, however, cover crops are essential. Good growth from well-inoculated legume cover crop seed (inoculated with rhizobia, a nitrogen-fixing bacterium) will frequently add 100 to 200 pounds of nitrogen per acre to the soil. Cover crops grown for their nitrogen benefit are often called green manure crops. The amount of organic matter produced, the quality of the organic matter, the ease of incorporation, and the amount of residue left in the soil may affect subsequent cultural operations. These are governed by the type of cover crop you choose, the inoculation used, the planting date, and the time at which you disk the cover crop in.

PEST MANAGEMENT

Throughout history, agricultural pests have been a problem for farmers, causing damage to seeds, seedlings, mature crops, and stored products. Any unwanted plant, insect, animal, or microscopic organism can be considered a pest if it somehow hinders production of a desirable crop. Pest management has always demanded a significant amount of a farmer's time and effort.

Different pests can affect almost every plant or animal. Sometimes their effect causes little or no damage. Growers should be aware of potential problems, however, because many can be controlled by prevention. Moreover, growers should be aware of as many of the various cultural, biological, and chemical ways of dealing with these problems as possible.

Insect and disease management manuals for many crops offer complete information on pest life cycles, avoidance techniques, and registered pesticides that are available for use in both conventional and organic farming systems. The University of California Integrated Pest Management (IPM) Program offers detailed information on pests of important fruit, nut, vegetable, and agronomic crops in California and specific management practices to control them. The information is accessible online (http://www.ipm.ucdavis.edu/PMG/crops-agriculture.html), or you can contact your local UC Cooperative Extension office for assistance in obtaining printed copies.

Certain pests are major problems of both annual and perennial crops every year, such as powdery mildew on grapes, brown rot on stone fruits, flea beetles on summer brassicas (e.g., broccoli, cabbage, cauliflower), or codling moth on apples and pears. These pests can destroy 100 percent of the marketable crop under certain conditions. In some cases the pest may damage the crop and also serve as a vector of a viral or bacterial disease. Whenever possible, choose a growing site that is less likely to experience devastating pest problems, such as a well-drained, isolated location with good air movement, and select varieties and rootstocks that are resistant to pests.

Arthropods and other pests. Examples of arthropod and insect pests are beetles, aphids, and caterpillars. There are other small-statured pests that are not arthropods, such as snails, slugs, and nematodes.Pests in this category can fly, crawl, or be blown or otherwise transported into crops.

Diseases. The pathogens that cause microbial diseases such as viruses, bacteria, and fungi are in the environment and may infest plants when conditions are right or when they are introduced by an insect pest.

Vertebrates. Vertebrate pests include rats, mice, deer, gophers, coyotes, rabbits, and birds.

Weeds. Weeds are another type of pest—one that is often challenging and costly to manage, especially in organic farming. Most soils contain weed seeds that can remain dormant for years until appropriate conditions exist for their germination. Many noxious perennial weeds can only be eliminated before planting. These include nutsedge, umbrella sedge, Bermudagrass, Johnsongrass, bindweed, wild blackberry, poison oak, and sheep sorrel.

Integrated Pest Management

Integrated pest management (IPM) is an approach that combines several techniques for dealing with pest problems. IPM may reduce your need for pesticides by incorporating various practices, such as crop rotations, scouting to see what pests are currently in your field, weather monitoring, use of resistant cultivars, timing of planting, disruption of pest mating with pheromones, and biological control of pests.

Biological control is the use of natural enemies to control pests at acceptable levels. Although biological control agents occur naturally in the environment, growers often purchase them to augment the population on their crops for pest control. This involves releasing increased numbers of already occurring predators and parasites at the most vulnerable period in the pest's life cycle. There are many commercial insectaries that provide biological control agents, and the use of this approach is expanding every year. To ensure success, users of commercially available predators and parasites need to understand the ecology of their local environment, including the target pest population and the predator or parasite to be released.

One of the essential components of IPM is a good knowledge of the pest. The effective use of IPM depends on your knowing the life cycle, ecology, and physiology of the pest so you will be able to intercept it at its most vulnerable stage. Pest species and population density must be monitored closely for you to make sound decisions on when and how to proceed. Preservation of beneficial organisms is important, and IPM also relies on other substitutes for conventional pesticides, such as traps and barriers, pesticides of low toxicity such as soaps, oils, and microbials, and changes in planting, irrigation, or cultivation procedures.

Pest Management in Organic Farming

In organic farming systems, you do have alternatives to conventional pesticides. A number of pest control materials have been approved for use by certified organic growers. For details of restrictions on organic pest management options, contact the California Department of Food and Agriculture's (CDFA's) Organic Program at 1220 N Street, Sacramento, CA 95814, (916) 445-2180, or visit their Web site (http://www.cdfa.ca.gov/is/i_&_c/organic.html), the USDA National Organic Program (http://www.ams.usda.gov/AMSv1.0/nop), and your accredited certifying agent.

A series of publications are available from the UC Division of Agriculture and Natural Resources on different aspects of organic vegetable production, including pest management. These publications, numbered consecutively from 7249 to 7254, are available at http://anrcatalog.ucdavis.edu or from your local UC Cooperative Extension office.

Pesticide Use Regulations

Agricultural pesticides—those used in both conventional and organic farming systems—may be used only for crops and pests for which toxicological research has resulted in official registration with both the U.S. Environmental Protection Agency (US EPA) and California's own state regulatory agency, CAL EPA. Most chemicals are registered for major crops grown in substantial quantities throughout a region. Because research is expensive, registration is frequently lacking for minor crops and fewer pesticides are available.

California has the most stringent pesticide regulations in the nation. Every farmer who uses pesticides is required to know, understand, and follow the pesticides laws and regulations. There are federal, state, and county pesticide regulations in effect in California.

Instructions on pesticide labels must be followed. The label is a legally binding document and the prestcide cannot be used in any way other than those that are specified on the label. If you fail to obey the instructions on the label, you are violating the law and may be subject to severe legal penalties. County agricultural commissioners enforce the program. All pesticide applications must be reported on a monthly basis to the County Agricultural Commissioner's Office.

It is often difficult to find effective pesticides registered for minor crops. To the USDA or pesticide manufacturers, even almonds, tomatoes, lettuce, tree fruits, grapes, and citrus are considered minor crops! Major chemical companies produce pesticides for major crops such as small grains, corn, soybeans, and cotton and register them with government regulatory agencies. In many cases, registering a pesticide for a minor crop is not economically justifiable for the companies. The federal government has established the Inter-Regional Project Number 4 (IR-4) to assist farmers with minor crop pesticide registration problems. There is an IR-4 office at the University of California, Davis.

All pesticides that have been registered in the United States are registered for either general or restricted use. General use pesticides are available to anyone. They are sold in supermarkets and hardware stores and are deemed to present minimal risk to the user or the environment when label directions are followed. No permits are required for general users (e.g., homeowners), but agricultural use, even of general use pesticides, must be reported monthly to the County Agricultural Commissioner's Office.

Restricted use pesticides are more hazardous and not available to the general public because they present potential health or environmental hazards. They are available only to users who have a pesticide use permit. To be granted a pesticide use permit you must pass a state examination and obtain the permit from your county agricultural commissioner. A permit may include conditions and restrictions beyond those on the pesticide label, such as those involving potential environmental impacts to groundwater, wildlife, and people.

Everyone who uses pesticides for agricultural purposes must first receive training. The county agricultural commissioner requires the farmer-trainer to keep a written record of what training has been given. Pesticide safety and training manuals are available through your local UC Cooperative Extension office or online through the UC ANR publications catalog (http://anrcatalog.ucdavis.edu).

As long as you follow label instructions, you can keep adverse environmental impacts to a minimum. The less pesticide material that gets into the environment, the less pesticide there is to cause environmental disruptions. Environmental disruptions are less likely when pesticides and other chemicals are used wisely and judiciously.

For additional information, contact your local county agricultural commissioner, UC Cooperative Extension Farm Advisor, agricultural supply company, licensed pest control adviser, or certified pesticide applicator.

"PEST" AND "PESTICIDE" DEFINED

Pests are organisms that attack, compete with, or adversely affect crops, humans, homes, pets, and livestock. Examples are insects, weeds, nematodes, parasites, fungi, snails, slugs, rats, mice, and animal and plant diseases. Any unwanted plant, animal, or microscopic organism may be considered a pest.

Pesticides are substances used to control pests. Included are insecticides, herbicides, defoliants, fungicides, nematicides, and rodenticides. Even a common substance like water, soap, oil, or salt may be considered a pesticide if it is used to control a pest.

CROP DIVERSIFICATION AND ROTATION

Farmers who manage only a few acres often have highly diversified operations. During a single growing season they may raise a variety of vegetables and fruits, often flowers and ornamentals, and sometimes poultry and livestock.

Crop Diversification

Crop diversification has two significant potential benefits for farmers: productivity and economic stability. Diversification can enhance economic stability by spreading production, financial, and market risks over a greater variety of farm enterprises. Signals from the marketplace should guide the grower to make a careful selection of crops and market niches. Ideally, the crop mix should be complementary; that is, all cultural practices should be performed with little competition for labor, equipment, or management time. An additional advantage to a more diverse crop mix is that it requires work more steadily throughout the year, permitting the grower to maintain a more stable workforce. This is not always possible because of factors that are beyond a farmer's control, such as unusual weather conditions or pest infestations. A good mix of crops can help prevent pests from becoming numerous enough to cause significant crop loss. One crop may be destroyed under such circumstances, but other crops may remain unaffected.

Crop Rotation

Crop rotation is the successive planting of different crops (usually two to five different crops) in the same field over time. Rotations also make possible the use of different tillage practices in a given field. Growers rotate a variety of crops, often including a cover crop or green manure crop. A successful rotation maintains fertility, helps manage pest and disease problems, reduces weeds, helps reduce soil erosion, and provides adequate, steady income. Crop rotation recycles nutrients, breaks pest cycles, and helps maintain a balance between soil organic matter accumulation and decomposition. Crop rotation also generally brings with it changes in tillage and other cultural practices.

A number of different rotation schemes are feasible. Additional information on diversification, including a crop rotation scheme, appears in chapter 5 of this book, Enterprise Selection.

GROWING VEGETABLES

Growing vegetables efficiently and productively in California requires year-round planning and scheduling. This includes production activities such as planting, fertilizing, irrigating, weeding, and harvesting. The following is a brief discussion of vegetable systems in California.

Field Selection and Practices

It is important to match the crop to the overall agro-climatic conditions of the area. Cool-season type vegetables, for example, including broccoli, cauliflower, celery, lettuces, and other leafy greens, tend to grow best in the mild, cooler valleys along the coast. The marine influence keeps temperatures relatively warm in winter and relatively cool in summer. Cool-season crops can also be grown successfully during fall, winter, and spring months in many of the hotter inland valley and desert areas.

Conversely, tomatoes, peppers, squash, and other warm-season vegetables are most widely grown in the hotter, inland areas. Some of these warm-season crops, such as peppers, may be grown in cooler, coastal areas during the warm summer months, while others such as tomatoes will not do so well under those conditions: while there is sufficient heat in coastal fields to produce and ripen tomatoes, the marine influence can bring early morning fog and dew to the fields during the summer months, aggravating fungal disease problems.

Soil type and water availability also influence site selection and how successfully a crop can be grown at a given site. The availability of sufficient water, as discussed previously in the irrigation section, will play a key role in the grower's selection of the crops and crop mixes for specific growing situations. Generally speaking, lighter-textured soils—with a higher percentage of sand and silt—are easier to work into a seedbed for vegetable planting and are often preferred for vegetable growing. Vegetables can be grown on heavier clay soils, too, but working these types of soil requires a more experienced farmer. Heavier clays also take more time to dry out following a rain before a field can be worked.

Planting and growing vegetable crops on raised beds is a good choice for an array of crops as well as for a range of soil types. The raised beds improve water drainage and reduce humidity around the growing crop. A bed configuration (its height and width) that is not too narrow or too wide enables the grower to plant and maintain diverse crops in the same field at the same time. This permits small-scale farmers to produce a limited number of rows of each crop in a field, simultaneously diversifying their products while growing relatively small volumes of each product. For example, a 40-inch bed width (center to center) would accommodate the planting on each bed of multiple rows of leafy greens, four or five rows of green beans of different types, two rows of peppers, one row of tomatoes, one row of sunflowers, and four rows of statice—all in the same field at the same time. Raised beds create excellent growing conditions and minimize potential problems from poor drainage on heavy or poorly prepared soils. Tillage equipment enables growers to break the bed down, incorporate residues and compost, and shape the bed up again for the next planting. A cultivator has to be readjusted to allow cultivation of the individual crops with different row spacings, but once the adjustment is made all crops with a similar configuration can be cultivated until the cultivator is adjusted for the next crop.

In some crops, such as bell peppers, cucumbers, or strawberries, you can use plastic mulch to cover the bed. The mulch improves moisture distribution uniformity and keeps soil off the fruit and plants. Depending on the color of the plastic mulch, it may help with soil heating or cooling as well. In many cases, you will need to fumigate the soil prior to mulching in order to prevent weed growth. Weeds will most certainly grow under white or clear mulch in the absence of fumigation, and black or clear mulch will cause soil heating in areas with hot summers. Some effective fumigants, including methyl bromide, are being phased out because of environmental concerns. Fumigation is not an option for organic growers, but many organic strawberry growers still use black plastic mulch to keep their fruit clean and keep soil from splashing onto plants and fruit. Growers will need to balance the costs and benefits of mulching against the limitations of fumigation options.

Row Crop Covers

Cool temperatures play a large role in adding to the uncertainties of vegetable production. Frosts and freezes can shorten the marketing period for many vegetable crops. Similarly, production of warm-season crops in many areas is severely restricted when cool temperatures occur during the growing season. By protecting plants with row covers, you can prevent losses from untimely frosts and freezes and provide a means to modify the environment around the plant favorably, resulting in more rapid growth, earlier maturity, and, possibly, increased yields.

High Tunnels

More expensive but longer lasting than row covers are high tunnels. The high tunnel is a simplified growing system that enhances crop growth, yield, and quality. Similar to a heated greenhouse, high tunnels are used to extend the growing season in the spring and fall. They are useful with a variety of crops. As a simplified growing system, however, they do lack the electrical components and automation you will find in a conventional greenhouse. Basic components of a high tunnel are a metal Quonset-hut-shaped frame, a single layer of polyethylene cover, and an irrigation system. Crops are grown in the ground, yet are protected from temperature fluctuations and light frost. Other benefits are reduced fluctuations in moisture, wind protection, soil warming, and often a reduction in pesticide usage due to a warmer, drier environment and the ability to use biological pest control.

High tunnels are relatively inexpensive compared to other types of greenhouse-style controlled environments, and they allow protected vegetable production with a limited capital investment. The system is most appealing to direct marketers, who can take advantage of having out-of-season production that, if successful, can be sold at a premium price.

Many vegetable crops, such as tomatoes, cucumber, peppers, and broccoli, do very well in high tunnels. The selection of specific crops for high tunnel cultivation will depend largely on what marketing opportunities are available to the individual grower. Construction for a series of 24- to 30-foot-wide bays of 12-foot-high tunnels costs $10,000 to $15,000 per acre (2009 values), but actual costs will vary depending on size, configuration, and supplier.

The University of California Vegetable Research and Information Center at UC Davis offers a series of free publications that give detailed management information for specific vegetable crops. These are available at your local UC Cooperative Extension county office or online (http://anrcatalog.ucdavis.edu) as well as the UC Davis Vegetable Research and Information Center (http://vric.ucdavis.edu/).

GROWING TREE FRUITS AND NUTS

There are two periods in the life of a perennial crop like an apple: the establishment phase and the fruit production phase. Establishment includes selection of the variety to plant and planting the orchard. Will you need a low-chill variety, as is required for production along the southern California coast, or a late-flowering variety, as is often needed in the Sierra foothills to avoid damage from late frosts? Also important is the selection of the trees' rootstock. For every fruit and nut variety you have a variety of rootstocks to choose from. When you plant an Elberta peach, you are actually using a Nemagard rootstock that has been grafted to the Elberta variety because Nemagard is more resistant to nematodes than the Elberta peach's own roots. In some cases, the chosen rootstock has dwarfing characteristics, which will reduce the amount of pruning required later on in the mature tree.

After varietal and rootstock selection, irrigation installation is important to consider if you are planting a perennial evergreen, such as citrus or avocado. This is not so critical for deciduous trees, like apple, since you plant them in the winter when they are dormant and do not transpire water. In the case of evergreens, you plant in late winter or early spring, and once the tree is planted irrigation is essential. Avoid planting into wet soil to avoid compaction. Try to plant at a depth of rooting that is similar to that of the nursery. If you are not sure what the nursery depth was, it is always better to plant slightly more shallow to avoid asphyxiation of the crown roots. During establishment, the most important activity for the plant is tree growth. Production of fruit or nuts should not be a concern yet: you are letting the tree establish itself for future production.

In fruit- and nut-bearing orchards, yearly production activities include pruning and thinning, irrigation, fertilization, and managing the orchard floor, along with pest management and harvest.

Pruning and Thinning

In the orchard's early years, the main function of pruning is to establish the shape that the mature tree will eventually have so it will be able to carry sufficient fruit loads. Pruning is done mainly in winter to balance vegetative vigor and good fruiting, but some additional pruning in the summer can enhance fruit color and reduce too-vigorous tree growth. Pruning lets the grower maintain a balance between currently fruiting wood and vegetative wood that will produce a crop in the future. Good pruning also reduce a tree's tendency toward alternate bearing (i.e., a large crop one year followed by little fruit the next, and so on) and tends to increase fruit size. Because pruning is labor intensive, it is costly. It also requires a great deal of knowledge and experience.

Fruit thinning improves fruit size and reduces alternate bearing. Thinning by hand is the most expensive operation in the orchard other than harvesting. Apples and olives are the only fruit for which chemical thinning with a growth regulator is commonly used. Chemical thinning of apples is often followed by hand-thinning. Some trees, such as nut crops, require no thinning.

Irrigation

Irrigation is quite possibly the most important production-related activity in a California orchard. It often determines the health of the tree, its growth rate, the number of fruit, and fruit size. Proper timing of irrigation is critical, as is the amount applied. When leafed out, young trees can only access a small amount of soil moisture and often require weekly or twice-weekly watering. As the tree grows, the irrigation schedule should change to reflect its larger rooting volume (which determines the amount of water the tree can access) and increased demand for water.

Fertilization

Fertilization must be scheduled according to recommended practices in your area. Leaf analysis is the best indicator of when to fertilize and what nutrients to apply. Adequate nutrients are particularly important for young trees in the establishment years to keep them healthy and optimize growth. Nitrogen, potassium, and some micronutrients are especially important when trees begin to bear a crop.

Managing the Orchard Floor

Cover crops are often grown on the orchard floor. If you follow this practice, choose a cover crop that will supplement fertility, help control erosion, and minimize additional water use. Cover crops such as grasses and low-growing, early maturing annuals will grow back from seed every fall. These are usually kept in check through periodic mowing. Other cover crops are planted every fall and then disked or tilled in in the spring as green manure. Since cover crops often compete for water with the orchard crop, growers with cover-cropped orchards normally use drip irrigation or microsprinklers to irrigate their trees. A winter cover crop that grows on rainfall alone and dies back when the soil moisture dries is one way to get a cover growing without applying any supplemental water. A winter cover will lower air temperatures somewhat, making frost problems somewhat more likely for evergreens such as citrus. All cover crops can harbor gophers, so trapping becomes more critical in a cover-cropped orchard than an orchard that is cultivated.

Hedgerows and Trellises

Fruit crops are commonly grown in hedgerows on size-controlled rootstocks; the small trees are generally more economical to farm. These trees can be pruned, thinned, sprayed, and harvested from the ground without the use of ladders. Fruit quality is often better and return on investment comes much sooner. The initial investment is higher, however, both to build a trellis system and to plant more trees per acre than in a traditional orchard. Many tree crops that used to be planted 20 feet by 20 feet apart on a 5-year system are now planted at a spacing of 10 feet by 16 feet or even closer on dwarfing rootstocks.

Investigate the many types of trellis and pruning systems in use and choose one that fits your farming methods and financial position. Apples in California, for example, have shown little benefit from the extra expense and greater labor required to attach trees to a trellis system.

GROWING SMALL FRUITS AND BERRIES

Cultivation of small fruits, as a group, uses many of the same critical cultural practices as other fruit and nut crops but often requires additional special management and practices such as trellising to maintain consistent production of high yields and high-quality fruit. Growing small fruits and berries such as grape, strawberry, blueberry, blackberry, and raspberry can be highly profitable for the space the crops occupy, but they are also labor intensive. Smaller fruit take more time to harvest than larger fruit such as apples or peaches. Berries are also highly perishable. Raspberries, for example, are among the most perishable of all fruit crops, with a typical shelf life of 7 to 10 days with optimum management. Berry crops require critical postharvest procedures, with careful handling and rapid precooling, for highest production and fewest losses.

Field Selection and Practices

Small fruits benefit from careful field selection. The best fields are those with sunlight all day long and well-drained and -tilled soils. Wind-sheltered areas are also preferred, especially for caneberries—blackberries and raspberries—to help minimize fruit lesions, blowing leaves and stems, and stress to the plants. All of the small fruit berries benefit from planting on raised beds, and growers often incorporate preplant fertilizer such as compost or other fertilizers as part of bed formation. Small fruits that are planted in properly prepared soil and are well maintained should have relatively few pest problems. However, fields with known disease problems should be avoided or should be treated with a soil fumigant. Keep the plants appropriately irrigated and fertilized and avoid conditions that expose flowers or fruit to frequent wetting by dew. Remove diseased fruit from the field to avoid associated pest problems. Organic growers will want to consider the use of cover crops, but more importantly they will want to select soils that have not been planted to crops that host pathogens that are harmful to small fruits.

Grapes

In California, grape production is often associated with winemaking. For the small grower, wine grape production can be a particular challenge because special licenses and permits are required if you want to make and sell wine. For this reason, many small growers decide to produce grapes for the fresh market or the dried market (as raisins). Many grape varieties are dual-purpose and can easily be grown for both fresh and dried markets. The vines are typically grown on trellises. That adds to their planting cost but makes it easier to prune and manage the crop for optimal yield. Depending on variety and climate, grapes may require special management for diseases, such as powdery mildew. In southern California, Pierce's disease, a deadly bacterial infection, limits plantings of traditional grape varieties, but some varieties are more resistant to the disease.

Strawberries

Raised beds help keep strawberry roots away from excessive moisture and offer growers a better medium in which to adjust soil conditions. In California, planting is usually from September through late November. Strawberries are generally planted 12 to 16 inches apart within and between rows, depending on the expected plant size. Strawberries are planted exclusively from crowns and should be grown from clean (certified) plant stock.

Caneberries

In California, raspberries are often planted from December through May, but in mild coastal areas you can plant them anytime. Raspberry plants are purchased as roots or canes. Root pieces are planted as a solid line into furrows, while canes are placed in evenly spaced holes, generally every 1 to 3 feet in rows 7 to 10 feet apart. Either way, raspberries eventually grow into a solid hedgerow (suckers grow to fill the spaces between initial cane-planting holes). Blackberries are usually planted from November to May, whether as transplants or as canes. Blackberries are spaced 3 feet apart within rows, with 6 to 10 feet between rows. They will retain their initial plant spacing throughout the life of the planting. Berry plants should be irrigated immediately after planting.

Caneberry Pruning

Two types of raspberry are planted in California: floricane fruiting and primocane fruiting. The floricane types are pruned 3 to 4 feet tall after one season of growth and bear fruit on branches that grow in the second season. The primocane types flower on the ends of first-year canes and as branches from floricanes, so they can be managed with more flexibility. Primocanes can be mowed to the ground in the fall or winter if you only want to pick the primocane fruit the following season, or they can be harvested first in the late summer and fall from the tips of primocanes and then pruned as if they were floricane types and picked the second year from branching floricanes. After that, the canes are cut to the ground so newly emerging primocanes can replace them, starting the cycle all over again.

Blueberries

Blueberry production is possible in many parts of California, provided that the grower selects the appropriate variety for the growing conditions and carefully follows soil management and cultural practices that are specific for blueberries.

Blueberry varieties vary dramatically in their growing requirements and fruiting characteristics. Highbush blueberries have the most desirable consumer appeal characteristics. Northern highbush types are for areas that receive more winter cold. Southern highbush types are best for milder southern and coastal areas. Many different commercial varieties are available within these two classes, and the selection of a variety that best suits specific local growing conditions is crucial with blueberries.

Blueberries also have specific soil requirements. The soil where blueberries are grown should be acidified to a pH of 5.0 or less; the water used for irrigating blueberries should also be acidified. Many soils in California have a pH of 7.0 to 7.5 or more, so acidification should be performed prior to planting. Blueberries are also intolerant of drought and of poor drainage. All of these factors taken together mean that soil cultural practices are very important for successful blueberry production.

The University of California Fruit and Nut Research and Information Center at UC Davis offers free publications via the Internet on the detailed management of specific fruit and nut crops (http://fruitsandnuts.ucdavis.edu). You can also contact your local UC Cooperative Extension office for information on these crops.

GROWING CUT FLOWERS, HERBS, AND ORNAMENTALS

On a small farm, cut flowers—annuals such as sunflowers, or perennials such as statice, lilies, or alstromeria—take up a small space and can supplement income from other crops. Growers can produce a mix of cut flower crops on a series of raised beds, and with repeated calendarized planting and transplanting they can offer a mix of cut flower products as well as premade bouquets. Herbs such as mint or lavender also can be worked into a small farming operation without occupying too much growing space, creating a more diverse and stable product offering. Marketing a mix of culinary, medicinal, and cosmetic herbs can be challenging, however, and a grower should have a clear marketing and sales plan developed before planting large areas of these crops. As with other vegetable, fruit, and ornamental crops, the grower should look to clear market signals and market research as guides to production quantities and timing.

Cut Flowers and Foliage

Cut flowers and foliage are usually grown in field beds similar to those used for vegetable crops. Sometimes—with small plantings of flowering bulbs, for instance—bulbs are planted in the container media that fills the plastic bulb crates in which the bulbs were shipped. Also, as in vegetable production, growers can use simple polyethylene-covered structures to enhance environmental conditions for flower growth and help reduce diseases that might occur in rainy periods. Permanent greenhouse structures, which control the plant environment to a greater extent, are sometimes used in cut flower and foliage production. Permanent greenhouses can greatly help manage the crop's growth and development and improve the quality of cut flowers. They can also allow you to grow certain cut flowers for year-round production, which may be necessary to satisfy markets. With increased control, though, comes a concomitant increase in costs for electricity, heating, and maintenance requirements. Labor requirements for producing a cut flower crop can be higher than for other, more mechanized crops because planting, pinching, disbudding, harvesting, and bunching are almost always done by hand.

Cut flowers and foliage can be annuals, biennials, or perennials. They are usually harvested for the fresh market and are extremely perishable. Once harvested, they must be sold quickly or kept in a cooler. Most cut flowers can benefit from postharvest cold storage temperatures as low as 33° or 34°F. Only semitropical or tropical cut flowers should not be stored at these cold temperatures. Many cut flowers are sensitive to ethylene gas, which is sometimes produced by fruits and vegetables, other flowers, or the crop itself. These ethylene-sensitive cut flowers can benefit from a relatively simple chemical treatment that reduces this sensitivity and greatly enhances the flowers' vase life for the consumer. Cut flowers and foliage can be produced for drying and are sometimes processed with glycerine to maintain aesthetic qualities once dried.

Ornamentals

Another option for the small farm producer is to grow potted ornamental plants. Possibilities include florist-quality flowering plants or foliage, bedding plants, flowering perennials, and woody landscape plants. Depending on the crop and final plant size, production time can be as short as just a few weeks, as with bedding plant transplants, or as long as several years, as with 15-gallon-container or larger landscape trees. Like cut flowers or foliage, a simple cover or greenhouse can speed production time and help prevent diseases in rainy weather. Sometimes greenhouses are required to produce florist-quality results for certain crops such as poinsettias and Easter lilies.

Potted ornamental plants can be grown in small containers—for example, as transplant 6-packs for bedding plants or in 1-, 5-, and 15-gallon plastic pots for flowering perennials or woody ornamentals. Particular care must be taken in the selection of growing media for the container to ensure that it meets the specific soil pH, water, and air requirements of the plant species. Soil media is often composed of soilless components such as fir bark, peat moss, coconut husk, sand, perlite, and vermiculite. The proper formulation of these components and the proper addition of mineral nutrients and amendments are what make a good medium. Make sure to obtain it from a reputable company. You can add fertilizers either as slow-release fertilizers or by fertigation during production as the crop uses up the initial charge of fertilizer in the container soil. This is a very competitive market area with large nurseries that can drive prices quite low. The small grower will therefore need to research the market before deciding what to grow.

GROWING SPECIALTY PRODUCTS

As a small farmer you may have limited resources, whether that means land, labor, finances, or experience. But you still want to make a living. The solution for many operations is to produce specialty products. Specialty products are crops that may have limited consumer demand and market potential but may also garner relatively high prices. If you decide to grow specialty crops, try something few or no other farmers are growing in your area and look for markets that are not yet saturated. Once there is plenty of any given product, it is no longer special. Growing a specialty crop will require experimentation both in the field and in the marketplace. Your local UC Cooperative Extension Farm Advisor may be able to give you information about a more common crop that is similar to the one you are testing. Library or Internet research and communication with universities or research centers located where the crop is commonly grown may help you gather information. You may find that you are a pioneer in researching growing techniques and developing the market for a new product.

The USDA Agricultural Marketing Service (AMS) Market News Service reports historical prices and volumes for many fruit, vegetable, ornamental, and other crops in diverse U.S. and foreign markets. USDA AMS information is available at http://www.ams.usda.gov/AMSv1.0/fv. You can also contact the USDA AMS by telephone (202) 720-2745, fax (202) 720-0547, or mail, 1400 Independence Ave. SW, Room 2503-S, Stop Code 0238, Washington, DC 20250-0238 for additional information and printed copies of reports.

It is difficult to develop a market for a new crop or specialty product. There is considerable time and risk because market potential has not been tested. See chapter 5, Enterprise Selection, and chapter 7, Marketing and Product Sales, for additional guidance.

New, Unusual, or Exotic Produce

Consumers like to try new varieties. A new color, flavor, size, or shape is appealing. A new item may be an unusual variety of a crop as common as lettuce. Lettuces are now available in many shapes, colors, and sizes, with names such as Red Oak Leaf, Rouge d'Hiver, and Black-Seeded Simpson. Many specialty crops are of European, Asian, or Latin American origin, having been introduced to the United States by immigrants, and many have been here for a long time but confined to ethnic markets. Examples are Belgian endive, bitter melon, lemongrass, specialty eggplants, chrysanthemum greens, nopales, and tomatillo. Perennial crops such as dragon fruit, longan, or lychee are also relatively new to California growers; however, prices are typically high in the marketplace at certain times of the year, and these crops may offer promising alternatives for growers in areas where these crops can be commercially produced.

Extended-Season Fruits and Vegetables

Demand is always high for tomatoes in January or for the first apples of the fall season. Maybe, through research or experimentation, you will discover a variety that matures much earlier or much later than what is currently available. Special cultural practices such as planting on south-facing slopes or using raised beds and sandy soils that warm more quickly can also allow you to shift production into an earlier harvest period. You may want to construct greenhouses or use plastic row covers to extend the growing season. Growers sometimes have to target windows of high prices between normal domestic production periods and the arrival of products from foreign production areas in order to garner higher market prices and improve profitability.

PROTECTED CROPPING

The main purposes of protected cropping (PC) are to accelerate or extend the crop season of vegetable, fruit, ornamental, or other specialty crops and to protect them from adverse environmental conditions and pests, especially when produced out of their regular season. A PC strategy can be a relatively simple, low-intensity, and lower-investment structure that simply protects the crop or extends its season. Protected cropping practices also include high-investment, sophisticated processes that attempt to control many aspects of the cropping environment. In an intensive PC effort, every aspect of a crop's environment, such as irrigation, nutrient uptake, temperature, lighting, carbon dioxide atmosphere, and humidity is analyzed and adjusted as needed.

Simple forms of PC include the plastic mulched beds, row covers, and high tunnels that we discussed earlier. More sophisticated, detailed, and intensive PC practices can require substantial labor or computer-controlled processes and a high level of capital input. The maintenance, crop production, and handling processes require constant vigilance and labor. On average, every 4,000 square feet of PC requires about 25 to 30 hours of crop care and upkeep per week. Access to precise costs and market analysis figures is especially important when you are developing this kind of project.

Protected cropping has important uses in California. As more and more prime agricultural land is developed for housing or industrial use, PC alternatives may be useful where agricultural production is forced onto marginal lands or less desirable locations. Prime agricultural lands located near urban areas are very expensive and carry high water and energy costs. Protected cropping offers potential solutions to agricultural production problems in both of these cases, but a detailed analysis of specific costs and anticipated income is critical.

Structure Types

Protected cropping accommodates several types of structures. The more commonly known are greenhouses (also called glasshouses). Traditionally, a greenhouse is a structure with walls and a roof of glass and a heating and cooling system. A greenhouse can also be made of thermoplastic such as Plexiglas, or it can be made up of frames covered with one or two layers of plastic, with or without auxiliary heating, that are used year-round or for only a few months every year.

Another type of structure is semicontrolled environment or protected shelter structure such as a shade house, cold frame, high tunnel (hoophouse), or low tunnel. These structures often employ the use of plastics, and their use is sometimes referred to as plasticulture. Unlike a greenhouse, a plasticulture structure does not control all environmental variables. The use of plastic in horticultural crop production has dramatically increased nationally and worldwide in the past 3 decades, via drip irrigation, plastic mulches, and high tunnels, for example. More and more small farmers are seeing different types of PC structures as essential to their farm operations. The type of structure is determined by the specific crop, availability of capital and labor, and by management capability and market strategy. Very specialized PC structures such as mushroom production houses can create the possibility for year-round production of certain specialty crops.

Heating and Cooling

The cost of heating and cooling structures, depending on system and location, can make PC an expensive business enterprise. In California, winters are relatively mild, with sunlight often abundant during winter and summer. Farmers may need to both heat and cool their PC structures depending on their crop and production area. Heating, cooling, and watering systems must be maintained and routinely serviced. In addition, contingency plans and backup systems must be in place in case any of these major systems should break down during periods of temperature extremes. Even a few hours' system failure when conditions are either too hot or too cool at a critical production period can result in complete crop failure. Running a PC enterprise can be either a frustrating or a rewarding experience, depending upon critical design or operation circumstances.

Potential Benefits and Drawbacks

Besides having to pull together the essential agricultural production skills, capital, and labor necessary to build and maintain a protected cropping structure, farmers must identify or develop the markets necessary to make the enterprise economically viable. The prospective grower should prepare a detailed business plan that carefully considers all of the production and marketing factors and variables necessary for a successful enterprise. Chapter 6, Farm and Financial Management, and Chapter 7, Marketing and Product Sales, will help you in this effort. Protected cropping operations can be undertaken to diversify and improve farm income by taking advantage of crop production that, in general, has a higher yield per unit of area than field-grown crops and that may have quality advantages as well. Additional benefits include year-round production, extended income, retention of current customers, attraction of new customers, and higher prices at times of the year when the product is otherwise in short supply. You can also provide extended or year-round employment for skilled employees who might otherwise be lost to other jobs at the end of the outdoor growing season.

While PC crops can be more productive per unit of area than field-grown crops, they also have disadvantages related to their higher investment and maintenance costs and the specialized management they require. Other disadvantages include the lack of any real break in the yearly work cycle and possible problems and or costs associated with disposal of waste (especially plastic waste).

HARVEST AND POSTHARVEST MANAGEMENT

Most vegetable, fruit, and ornamental crops require hand harvesting and subsequent postharvest management that typically involves washing, sorting, packing, and cooling operations. New growers should become familiar with accepted industry standards for packing container sizes, product grades and presentation, and cooling, storage, and transportation practices. Postharvest management is as important as the various aspects of production management if you want to deliver high-quality product to the marketplace. Likewise, new growers need to find experienced marketers and marketing agents who know the crops and markets so they can be sure their products will move efficiently through the marketing stream with relatively few losses. Chapter 10, Postharvest Handling and Safety of Perishable Crops, offers more information and guidance.

RESOURCES AND REFERENCES

Cover Crops

Clark, A. 2007. Managing cover crops profitably. Third edition. Sustainable Agriculture Network Handbook Series Book 3. USDA SARE, Beltsville, MD. Pp. 244. http://www.sare.org/publications/covercrops/covercrops.pdf.

Ingels, C. 1998. Cover cropping in vineyards: A grower's handbook. University of California Division of Agriculture and Natural Resources, Oakland. Publication 3338. Pp. 162.

Miller, P. 1989. Covercrops for California agriculture. University of California Division of Agriculture and Natural Resources, Oakland. Publication 21471. Pp. 24.

Pest Management

Alitieri, M., C. Nicholls, and M. Fritz. 2005. Manage insects on your farm: A guide to ecological strategies. Sustainable Agriculture Network Handbook Series Book 7. USDA, SARE, Beltsville, MD. Pp. 130. http://www.sare.org/publications/insect/insect.pdf.

Bowman, G. 2001. Steel in the field: A farmer's guide to weed management tools. Sustainable Agriculture Network Handbook Series Book 2. USDA SARE, Beltsville, MD. Pp. 128. http://www.sare.org/publications/steel/steel.pdf.

Canevari, M. 1998. Pesticide safety for small farms: A grower's guide to pesticide safety. University of California Division of Agriculture and Natural Resources, Oakland. Publication 21555. Pp. 32.

Dreistadt, S. 2008. Integrated pest management for avocados. University of California Division of Agriculture and Natural Resources, Oakland. Publication 3503. Pp. 222. http://anrcatalog.ucdavis.edu/IPMManuals/3503.aspx.

———. 2008. Integrated pest management for floriculture and nurseries. University of California Division of Agriculture and Natural Resources, Oakland. Publication 3402. Pp. 422 http://anrcatalog.ucdavis.edu/IntegratedPestManagement/3402.aspx.

Flaherty, D. 1992. Grape pest management. Second edition. University of California, Division of Agriculture and Natural Resources, Oakland. Publication 3343. Pp. 412.

Flint, M. L. 1991. Integrated pest management for citrus. Second edition. University of California Division of Agriculture and Natural Resources, Oakland. Publication 3303. Pp. 144. http://anrcatalog.ucdavis.edu/IntegratedPestManagement/3303.aspx.

———. 1992. Integrated pest management for cole crops and lettuce. University of California Division of Agriculture and Natural Resources, Oakland. Publication 3307. Pp. 112.

———. 1998. Integrated pest management for tomatoes. Fourth edition. University of California Division of Agriculture and Natural Resources, Oakland. Publication 3274. Pp. 118. http://anrcatalog.ucdavis.edu/Tomatoes/3274.aspx.

———. 1998. Pests of the garden and small farm: A grower's guide to using less pesticide. Second edition. University of California Division of Agriculture and Natural Resources, Oakland. Publication 3332. Pp. 276. http://anrcatalog.ucdavis.edu/SmallFarms/3332.aspx.

———. 2002. Integrated pest management for almonds. Second edition. University of California Division of Agriculture and Natural Resources, Oakland. Publication 3308. Pp. 231. http://anrcatalog.ucdavis.edu/IntegratedPestManagement/3308.aspx.

Flint, M. L., and S. Driestadt. 1998. Natural enemies handbook. University of California Division of Agriculture and Natural Resources, Oakland. Publication 3386. Pp. 154. http://anrcatalog.ucdavis.edu/SustainableandOrganic/3386H.aspx.

Ohlendorf, B. 1999. Integrated pest management for apples and pears. Second edition. University of California Division of Agriculture and Natural Resources, Oakland. Publication 3340. Pp. 231. http://anrcatalog.ucdavis.edu/IntegratedPestManagement/3340.aspx.

Strand, L. 2006. Integrated pest management for potatoes in the Western U.S. University of California Division of Agriculture and Natural Resources, Oakland. Publication 3316/011. Pp. 167. http://anrcatalog.ucdavis.edu/VegetableCrops/3316.aspx.

———. 2008. Integrated pest management for strawberries. Second edition. University of California Division of Agriculture and Natural Resources, Oakland. Publication 3351. Pp. 176. http://anrcatalog.ucdavis.edu/IntegratedPestManagement/3351.aspx.

———. 2003. Integrated pest management for walnuts. Third edition. University of California Division of Agriculture and Natural Resources, Oakland. Publication 3270. Pp. 136. http://anrcatalog.ucdavis.edu/IntegratedPestManagement/3270.aspx.

UC IPM. 1990 (and periodically updated). UC IPM pest management guidelines. University of California Division of Agriculture and Natural Resources, Oakland. Publication 3339. Information on insect, mite, disease pests, nematodes, and weed pests for major crops. Also online: http://www.ipm.ucdavis.edu/PMG/crops-agriculture.html.

Soil and Fertility Management

Chaney, D. 1992. Organic soil amendments and fertilizers. University of California Division of Agriculture and Natural Resources, Oakland. Publication 21505. Pp. 32.

Gaskell, M. 2006. Soil fertility management for organic crops. University of California Division of Agriculture and Natural Resources, Oakland. Publication 7249. Pp. 8. http://anrcatalog.ucdavis.edu/pdf/7249.pdf.

Hanson, B., and L. Schwankl. 2001. Chemigation in tree and vine microirrigation systems. University of California Division of Agriculture and Natural Resources, Oakland. Publication 21599. Pp. 12.

Magdoff, F., and H. Van Es. 2009. Building better soils for better crops: Sustainable soil management. Third edition. Sustainable Agriculture Research and Education Handbook Series Book 10. USDA, SARE, Beltsville, MD. Pp. 314. http://www.sare.org/publications/bsbc/bsbc.pdf.

Storie, R. E. 1988. Generalized soil map of California. University of California Division of Agriculture and Natural Resources, Oakland. Publication 3327.

Specialty Crops

California Rare Fruit Growers. A nonprofit organization that fosters public interest in preserving and producing rare fruits. Publishes a bimonthly magazine. http://www.crfg.org/index.html.

Kowalchik, C., and W. H. Hylton, eds. 1998. Rodale's illustrated encyclopedia of herbs. Rodale Press, Emmaus, PA. Pp. 545.

Morgan, J. F. 1987. Fruits of warm climates. Florida Flair Books, Miami, FL. Pp. 505.

Myers, C., et al. 1998. Specialty and minor crops handbook. University of California Division of Agriculture and Natural Resources, Oakland. Publication 3346. Pp. 184.

Rubatzky, V. E., and M. Yamaguchi. 2007. World vegetables: Principle, production, and nutritive values. Second edition. Aspen Publishers, Inc., Gaithersburg, MD. Pp. 704.

Stamets, P. 2000. Growing gourmet and medicinal mushrooms. Third edition. Ten Speed Press, Berkeley, CA. Pp. 592.

Stamets, P. and J. S. Chilton. 1983. The mushroom cultivator: A practical guide to growing mushrooms at home. Agarikon Press, Olympia, WA. Pp. 415.

Stephens, J. M. 1988. Manual of minor vegetables. University of Florida, Gainesville, FL. Pp. 123.

Whealy, K. 2005. Garden seed inventory: Inventory of seed catalogs listing all non-hybrid vegetable seeds available in the United States and Canada. Sixth edition. Seed Saver Exchange, Decorah, IA. Pp. 504.

———. 2009. Fruit, berry, and nut inventory: An inventory of nursery catalogs and websites listing all fruit, berry, and nut varieties available by mail order in the United States. Fourth edition. Seed Savers Exchange, Decorah, IA. Pp. 384.

Water and Irrigation Management

Bowers, W. D. 1989. Water-holding characteristics of California soils. University of California Division of Agriculture and Natural Resources, Oakland. Publication 21463. Pp. 92.

Goldhamer, D. 1989. Drought irrigation strategies for deciduous orchards. University of California Division of Agriculture and Natural Resources, Oakland. Publication 21453. Pp. 15.

Hanson, B., L. Schwankl, and A. Fulton. 1999. Scheduling irrigations: When and how much. University of California Division of Agriculture and Natural Resources, Oakland. Publication 3396. Pp. 204. http://anrcatalog.ucdavis.edu/Irrigation/3396.aspx.

Other Resources and References

American Vegetable Grower, 37841 Euclid Ave., Willoughby, OH 44094. Monthly magazine on culture, management, equipment, etc. for vegetable growers. http://www.meistermedia.com/publications/vegetable.html.

Grubinger, V. P. 1999. Sustainable vegetable production from start-up to market. Natural Resource, Agriculture, and Engineering Service (NRAES), Ithaca, NY. Pp. 268.

Harrington, G. 1984. Grow your own Chinese vegetables. MacMillan Publishing, New York. Pp. 268

Kader, A. A., tech. ed. 2002. Postharvest technology of horticultural crops. Third edition. University of California Division of Agriculture and Natural Resources, Oakland. Publication 3311. Pp. 535. http://anrcatalog.ucdavis.edu/Postharvest/3311.aspx.

Maynard, D., and G. Hochmuth. 2007. Knott's handbook for vegetable growers. Fifth edition. Wiley, Hoboken, NJ. Pp. 640.

Mohler, C., and S. E. Johnson, eds. 2009. Crop rotation on organic farms: A planning manual. Sustainable Agriculture Research and Education Program and Natural Resource, Agriculture, and Engineering Service (NRAES), Ithaca, NY. Pp. 156. http://www.sare.org/publications/croprotation/croprotation.pdf.

Otto, H. W., R. Branson, and K. Tyler. 1990. Guide for fertilizing vegetables. In R. Voss, ed., Vegetable production. University of California Division of Agriculture and Natural Resources, Oakland. Publication ANRP012. Pp. 70.

Sutter, S. 1993 (revised). New federal pesticide safety standard issued. University of California Cooperative Extension, Fresno County, Fresno, CA.

———. 1993 (revised). Pesticide safety guide for agricultural workers. University of California Cooperative Extension, Fresno County, Fresno, CA.

Swiader, J. M., and G. W. Ware. 2001. Producing vegetable crops. Fifth edition. Prentice Hall, Upper Saddle River, NJ. Pp. 640.

United Fresh Fruit Produce Association, 1901 Pennsylvania Avenue NW, Suite 1100, Washington, DC 20006. Sells leaflets on more than 80 fresh fruits and vegetables, offering data on produce growing, shipping, and marketing, as well as historical and botanical information. http://www.unitedfresh.org.

10
Postharvest Handling and Safety of Perishable Crops

Trevor Suslow, Elizabeth Mitcham, and Marita Cantwell

Small farms, as mentioned in earlier chapters of this book, are usually defined on an economic basis as farms with an annual gross income of $250,000 or less (United States Department of Agriculture [USDA] ERS, 2007). It is also true, though, that small-scale farm management issues can correspond to a farm's small physical footprint, the balance of family versus nonfamily labor, and the grower's degree of control over how the agricultural output reaches the ultimate consumer. Smaller farms often have to operate differently than bigger operations because they have more limited access to harvest and postharvest management equipment and technologies and to primary marketing channels and destinations. This chapter will give you an overview of general postharvest handling considerations for fruits and vegetables, focusing primarily on practices applicable to small-scale management with moderate to low annual sales of these commodities. Details of modern postharvest technologies and long-distance transportation, which can be capital intensive, are available from a variety of resources. For more information on this, see the online Postharvest Yellowpages (http://postharvest.ucdavis.edu).

Besides the visual quality, sensory quality, and nutritive value of produce, any discussion of postharvest management of perishable horticultural foods that are eaten fresh must include an overview of food safety management. The need to minimize the potential for injury or illness to consumers is independent of a farm's scale or range of distribution and is the clear responsibility of primary producers and suppliers. Beyond the general reaffirmation of the need for commonsense measures to ensure food safety, this chapter gives some more specific guidance that can increase your awareness of preharvest risks and the need to plan for responsible postharvest management and meet the expectations of public health regulations (U.S. FDA, 2008).

PLANNING FOR POSTHARVEST QUALITY

The effort to achieve economic reward through the marketing of edible horticultural foods must begin well before harvest. Selection of the right seeds or perennial varieties or cultivars, including rootstocks, can be critical in determining the postharvest performance of any commodity. Individual cultivars vary in their inherent potential for firmness retention, uniformity of shape and characteristic color, disease and pest resistance, and shelf life in terms of visual quality and sensory quality (i.e., sugar:acid balance; aroma volatiles), to list just a few key traits. Small farm producers typically include or focus on specialty and heirloom cultivars and harvest strategies that maximize the sensory quality of produce, such as growing it to near full ripeness on the vine, in order to attract and satisfy direct-market consumer demand.

Both conventional and organic growers have recognized and responded to the expanding consumer demand for better flavor profiles in standard varieties as well as true heirloom and heirloomlike varieties. Heirloom varieties and those selected for novelty, ethnic culinary, or flavor traits are typically best suited for small-scale production and local marketing. Some heirloom or specialty varieties are not well adapted for growth in the arid Mediterranean climate of California and develop disorders or conditions in preharvest phases that become serious postharvest problems. Many varieties in this category simply do not hold up well to current postharvest handling and distribution methods. The main problems are bruising and cracking, compression damage in pallet loads, excessive softening, shriveling and wilting from water loss, and decay.

In addition to genetic traits, environmental factors such as soil type, temperature, wind during fruit set, frost, and rainy weather at harvest can adversely affect produce conformance with appearance (shape, uniformity, size) standards, storage life, suitability for shipping, and quality. Other local climatic factors such as seasonal high winds, fog, or low humidity during critical fruit development stages can result in deformities, poor shelf life, or accelerated decay compared to produce from other regions. In general, seasonally and regionally selected varieties, popular for domestic and export distribution markets, overcome most of these potential market defects, so small-scale growers often include these along with specialty varieties in an overall profitable production plan.

Cultural practices may have a dramatic impact on postharvest quality. For example, too little seedbed depth for carrots may result in sunburned shoulders and green cores in many of the specialty carrots favored most by consumers at farmers markets. As the blunt-end root penetrates into more-compacted soil, the crown pushes above the soil line where it is exposed to the sun. Carrots that develop green core tend to be bitter well beyond the margin of visible greening.

Management of the health and density of foliar parts of the carrot plant is also an important way to reduce the carrots' exposure to intense sunlight. Improper pruning, thinning, fertilization, and disease control can reduce produce quality. For example, inadequate pruning and thinning of cherry tomato plants results in excessive fruit splitting and higher rates of postharvest decay from infections that started in the field under high humidity, poor air circulation conditions.

Excessive nitrogen in the soil and inconsistent irrigation may result in a variety of quality disorders, such as thin walls and uneven ripening in sweet peppers and calcium deficiency disorders in many fruits. Quality loss also results from rough handling during and after harvest. Thus, planning, protection, and responsiveness to changing local conditions are vital, both in production and during postharvest handling, if you want to avoid immediate causes of deterioration, slow the deterioration of produce during short-term storage, and reduce losses in distribution channels.

HANDLING AT HARVEST

The inherent quality of produce cannot be improved after harvest—only maintained for the expected window of time (the shelf life) characteristic of the commodity and variety. Use commodity-specific maturity indices to determine when is the best time for harvest. While you can harvest some commodities successfully at a mature stage and let them ripen during postharvest handling (thereby improving their eating quality), the products' actual sensory and nutritive properties were fixed during preharvest production and captured at the moment of harvest.

Part of what makes for successful postharvest handling is an accurate knowledge of the length of this window of opportunity under the specific conditions of production, season, method of handling, temperature during short-term storage, distance to market, and the impacts of mixed-load handling. It is also critical to have a good understanding of how easy it is to miss this window as a result of poor planning and management. For a small farm, the grower may be more likely to harvest and market produce at or near peak ripeness than the owner of a larger operation would be.

As discussed above, small farm operations often select or include specialty varieties with shelf lives and shipping traits that are suboptimal or even inherently short. The operating principles described below are important in all operations but carry special importance for many small-scale producers who have less access to postharvest cooling equipment, covered and cooled grading, sorting, and packing areas, and refrigerated short-term storage and loading dock facilities. Resources for refrigerated transportation may also be lacking.

Field packing. Soft fruits like berries and many vegetables are packed directly in the field. This minimizes additional direct handling of the produce and can reduce the time between picking and cooling. The disadvantages are that quality control—primarily produce uniformity within a packed carton or unit—is more difficult to achieve in the field than in a packinghouse, since grading, sorting, and trimming are limited to whatever product is within immediate reach. Field uniformity, determined by variety selection and preharvest management, is critical for efficient field packing if uniformity of size and shape is an important condition of sale. Furthermore, since sorting and trimming must be done by hand, variability between workers may become an issue and necessitates close training and supervision in the field. In a field-pack operation, cleaning and application of postharvest chemicals to reduce water loss or prevent decay typically are not done. Another important consideration is that workers in the field may be exposed to less comfortable environmental conditions or even highly stressful conditions compared to those experienced by packing shed workers, and that can influence their ability to execute a quality pack.

Field packing can be as simple as a picker with a box, or you may use a harvest aid to increase worker comfort and efficiency. For strawberries, the picker places the fruit into baskets that are already loaded into a corrugated fiberboard tray. In many cases the tray is carried on a small wheeled dolly that fits easily within a single furrow. As a tray is filled, the picker takes it to a shaded holding area. Table grapes are similarly packed at the vine. The picker grades, trims, and bags the product and places the bags into boxes. Alternatively, the picker may accumulate grapes in a tote and bring the tote to a small, mobile field-side packing station that has a canopy to provide shade for both worker and product. In small operations close to a shipping facility, the time from picking to cooling can be less than 1 hour, which greatly benefits postharvest quality during transit to the final consumer.

Harvesting for packing in a central area. Most fresh market products are harvested by hand into buckets, lugs, baskets, or canvas bags, which are then emptied into field totes or larger bins for transport to a central packing area. In some operations, produce may be packed directly from the bucket or tote, which has been transported from the field on a flatbed trailer fitted with a racking structure. For highly perishable commodities—edible pod and sugar snap peas or young leafy greens, for example—quality is best maintained if you place harvest totes or lugs directly into cool conditions during harvest, such as a small cab-over refrigerated truck.

Care should always be taken for specialty crops or fruit harvested at a thin-skinned stage (e.g., specialty cucumbers) or ripe stage (e.g., apricots) to minimize abrasion from buckets and totes or sorting surfaces, including field grit, which can be abrasive. Compression injuries can result if the harvest containers are too deep. Metal or plastic picking buckets are typically used for the softer fruit such as cherries, and bottom-dump picking bags are used for fruit with less potential for compression bruising, such as citrus fruit or pears. The surfaces of the harvest container should be clean and smooth. All containers need periodic washing to remove any soil adhering to their interior walls and external surfaces, especially if they are designed for stacking and nesting. Harvest residues and juices may be an ideal location for growth and spore production of spoilage and decay organisms, which can then contaminate each harvested lot. We strongly recommend that you conduct preseason and periodic inspection of harvest containers, looking for splitting, chipping, splinters, and excessive abrasive areas.

During accumulation for centralized packing, produce can suffer physical injuries from pulling, improper clipping, fingernail cuts, grit adhering to gloves, scuffing on a gritty, worn, or poorly designed harvest-aid conveyor belt, dropping into picking buckets or bags, overfilled picking containers, striking of the containers (especially soft-sided bags) against limbs and ladders, pickers leaning against picking bags while picking, transferring of product into field lugs without due care, and overfilling of field containers. Even a short drop or tossing of a tote or lug onto a pallet can cause substantial impact bruising that is not always apparent until ripening or later stages during distribution when affected areas become discolored and show signs of advanced water loss or decay. Road vibrations and sharp shocks in a pickup, flatbed, or trailer can cause significant visible damage, even on a relatively short trip to a farmers market.

Temperature protection in the field. To maximize postharvest quality and marketability, reduce the time between produce harvest and cooling. Many products, such as strawberries, asparagus, leafy greens, and tender herbs, lose visible and measurable amounts of quality during just 1 or 2 hours of delay between harvest and cooling. Weather conditions during harvest strongly influence the length of this window of time. The characteristic respiration rate related to the rate of quality deterioration of the commodity is another important factor. (For an explanation of postharvest respiration and product quality, see *Postharvest Technology of Horticultural Crops*, ANR Publication 3311.) Frequent transfer of products to the cooler or a shaded packing area minimizes the amount of time they spend in conditions conducive to heating and respiration-related deterioration. If your harvest rate is slow, make several small trips rather than waiting until you have a full load of products to transport.

While in the field, shade products after they are harvested. Products left in the sun can warm considerably above the air temperature. Dark-colored products can warm even more quickly than light-colored products. In orchards and vineyards, you can place products in the shade of the trees or vines. If no natural shade is available, use portable shading to reduce exposure to the sun. Portable shading may be more effective than natural shading because it will not shift as much as natural shade will as the sun moves across the sky. Placing empty packages or lugs over the top of stacks of packages provides some protection. Shading keeps products from warming above the ambient air temperature, but even a mild breeze can cause harvested product in the shade to quickly warm to near the ambient air temperature. Especially in hotter climates, lugs and totes that are stacked and held for more than a few hours should have adequate venting to prevent low oxygen conditions that lead to off-flavors and increase spoilage.

During periods of high field temperatures, products should be harvested early in the day to reduce product warming. This also maintains product turgidity (firmness), size, and gloss, and reduces the amount and cost of cooling needed. Because of their high turgidity early in the day, some products with sensitive skin (such as citrus) or prone to splitting and bruising (such as tomatoes if turgid and cool) are purposely harvested later in the day to avoid injury to the skin. Sometimes harvesting must be delayed to allow condensation or dew to evaporate and so reduce postharvest skin staining and decay.

As a general approach, the following practices can help you maintain product quality:

- Harvest during the coolest time of day to maintain low product respiration and transpiration (water loss) rates.
- Minimize the extent of harvest cut or picking wounds, bruising, crushing, or damage from humans, equipment, or harvest containers.
- Shade the harvested product in the field to keep it cool. Covering harvest bins or totes with a foam-backed reflective pad greatly reduces heat gain from the sun, water loss, and premature senescence.
- If possible, move the harvested product into a cold storage facility or postharvest cooling treatment as soon as possible. For some commodities such as berries, tender greens, and leafy herbs, 1 hour in the sun is too long.
- Do not compromise high quality product by mingling it with damaged, decayed, or decay-prone product in a bulk or packed unit.
- Only use cleaned and, as necessary, sanitized packing or transport containers.

Preparation for Packing

Cleaning and washing. Some commodities—in particular, root and tuber vegetables and leafy greens grown close to the soil—may need or benefit from cleaning to remove soil, small adhering insects, and other contaminants. In small farm operations, the degree to which this kind of cleaning is necessary depends on the requirements or expectations of the specific market. Food-grade detergent washes and fruit and vegetable wash-aides (typically, approved plant extracts) are sometimes used along with soft brushes or sponges followed by clear water rinsing. Many peaches are wet brushed to remove the trichomes (peach fuzz). Oranges are sometimes washed with a high-pressure (350 psi) spray to remove scale insects and surface mold, while vine-ripe tomatoes and specialty melons may receive a gentle disinfectant water spray and be individually hand-rubbed with a clean cloth for better presentation.

Special operations. A wide range of special operations may be needed to prepare the products for final sorting. Unwanted leaves, stems, and roots are removed from some vegetables. Removal of the calyx or additional trimming of the stem may be done prior to packing for some fruits and fruitlike vegetables. Other products such as asparagus spears, celery, and green bunching onions are trimmed to a uniform length. For many high-end direct markets and value-added products, small farm enterprises are increasingly introducing specialty consumer packing and microwave-ready trays, which generally include or require trimming (of sweet corn, squash, or green beans, for instance) or careful brush washing (for new potatoes).

Disease control. Careful handling to prevent product damage and careful monitoring of storage conditions to maintain the lowest safe temperature for the product (see "Importance of Optimal Storage and Shipping Temperatures," later in this chapter) are the two most effective strategies for controlling postharvest diseases. For products that are washed or cooled with water, disinfection of both single-pass and recirculated water is an important preventive and control step. See "Sanitation and Water Disinfection," later in this chapter and also *Postharvest Chlorination: Basic Properties and Key Points for Effective Distribution* (ANR Publication 8003). Some postharvest disease control treatments may be applied during packing. Fungicide applications, if used, are commonly applied when the fruit is spread on conveyor belts or rollers, often immediately after washing. Fungicides are often incorporated into fruit waxes to help achieve a uniform surface application. Background and information on postharvest fungicides and alternative decay control strategies may be found in *Postharvest Technology of Horticultural Crops* (ANR Publication 3311) as well as from other sources specific to small farms and organic producers and handlers, including ATTRA National Sustainable Agriculture Information Service (http://attra.ncat.org/). All chemical applications must be made in strict conformity to government regulations.

Heat treatment, especially in the form of hot water treatment, has been shown to be effective for postharvest disease control of many products. It is widely used on papayas and oranges, typically before or at the start of packing, and has been found to be beneficial for tomatoes and melons. The use of gaseous ozone is increasing in cold storage facilities, including small farm operations, to reduce the build-up of mold spores and the transfer of decay from fruit to fruit in bins and packed cartons. See *Ozone Applications for Postharvest Disinfection of Edible Horticultural Crops* (ANR Publication 8133).

Sorting and Grading

Many growers hand-sort their produce to segregate it by maturity, color, size, and grade. Small fruit and vegetables require more sorting decisions per package than do larger products. Light levels of 500 to 1000 lux at the sorting surface are usually adequate; older workers may need twice as much light as their younger co-workers. The level of lighting in the sorting and grading area should be uniform: workers should not have surfaces in their field of view that vary in luminance (level of reflected light) by more than 3 to 1. Hand-sizing guides or rings and good-quality placards or photos of grades and defects are simple tools that help packers make quick and consistent sorting decisions. Machines are also available that sort fruit by size based on weight, volume, or electronic images. Fruit pass on a conveyor through the sizing equipment and are then packed by machine or delivered to a work station for final pack quality judgment.

Packing the Product

Packaging serves a number of functions including protection of the product from damage during transport, unitization, and advertising and display of the product at retail markets. Packaging performance issues include box strength under the conditions of use, location and degree of ventilation for product cooling, and container size and shape. Function and performance needs vary with the fruit or vegetable product. Other issues to consider include content of recycled material, capacity to be recycled, disposal fees associated with packaging materials, and suitability of containers for retail display. The reuse of corrugated fiberboard containers is strongly discouraged because of the risk of cross-contamination with pathogens or chemicals. Reuse of corrugated containers and trays is prohibited in some states and for certain commodities. Clean, sanitized, or new containers are strongly recommended for produce regardless of the scale of operation.

One-touch merchandising, where the packed produce is loaded directly into reusable, high-quality containers, shipped through existing supply chains, and merchandised at retail outlets with minimal handling, is becoming increasingly popular, even for smaller suppliers and handlers. Reusable plastic containers (RPCs) and high-quality display-sized fiberboard boxes have been established as fulfilling this need, but the substantial cost of the RPCs is typically borne by the shipper. There has been considerable growth in the use of consumer packages such as vented bags, netted bags or sleeves, and clamshell containers. These containers provide extra protection for products such as berries and cherries, maintaining high relative humidity and providing protection from mixing and spillage during retail display. One point of caution: conditions in clamshell containers and bags that reduce water loss may also promote growth of surface molds or decay pathogens at warmer storage and transport conditions.

A variety of netted plastic sleeves or sacks and bags with interlocking closures are especially popular for consumer value packs and ready-to-eat or ready-to-cook value-added produce. Packed bags are also used increasingly to help maintain the quality of whole fruit. Bags provide no protection from impact bruising, but they do reduce moisture loss by maintaining a high relative humidity around the product. Tight-filled bags can also reduce damage from high-frequency road vibrations during transport by preventing rolling abrasions. Bags can sometimes provide a modified atmosphere, depending on the type of polymer film used and how well the bag is sealed. Recommendations for use of modified atmospheres can be found online at the UC Postharvest Technology Research and Information Center's Web site (http://postharvest.ucdavis.edu/Producefacts).

Product is often hand-packed in small-scale operations. The goal is to pack a fixed count in trays or boxes and always immobilize the product within the package. The packer may also sort, grade, and trim the product before placing it in a box. Workers need adequate lighting, and their work space should be organized to reduce physical stress by minimizing the need for reaching (figure 10.1).

Figure 10.1. Components of a hand-pack operation.

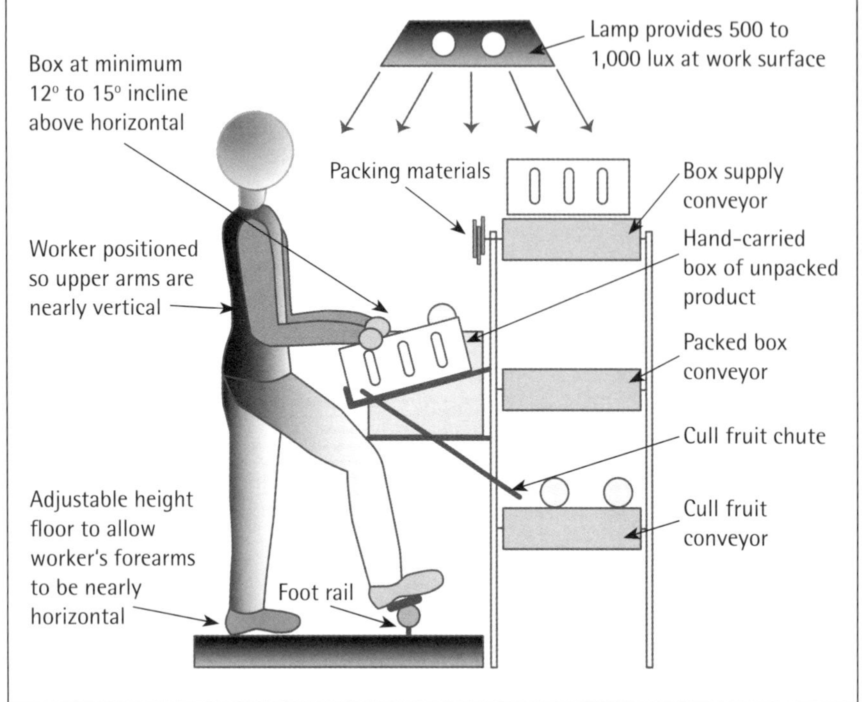

The machinery used in hand-packing is mainly for materials handling, not for sorting, grading, or box filling. Product is either hand-carried to the packer or a return-flow belt is used, where the packer selects product of a desired size from the belt. An efficient rate of delivery for product and packaging materials to the packers and prompt, efficient removal of filled packages are important to packing efficiency.

Reducing Water Loss

Temperature management plays a key role in limiting water loss in storage and transit. As the primary means of lowering respiration rates of fruits and vegetables, temperature has an important relationship to relative humidity and thus directly affects the product's rate of water loss. The relative humidity of ambient air conditions in relation to the relative humidity of the crop (essentially, 100%) directly influences the rate of water loss from produce at any point in the marketing chain. Water loss can be rapid even though the resulting postharvest damage may not be immediately visible. The effects of water loss are cumulative and may result in wilting, shriveling, loss of crispness, browning, stem separation, or other defects. For some commodities such as table grapes, strawberries, specialty peppers, and squash, as little as 2 percent water loss will drastically reduce their market attractiveness and consumer appeal.

Transport to and display at roadside stands or farmers markets often result in extended periods of exposure of sensitive produce to direct sun, warm (or even hot) temperatures, and low relative humidity. Rapid water loss under these conditions can result in limp, flaccid greens and a loss of appealing natural sheen or gloss in fruits and vegetables. By providing postharvest cooling before and during transport and a shading structure during display, you can minimize rapid water loss at these market outlets.

Fruit and vegetable waxes are effective at reducing water loss and enhancing product appearance, but it is important that the produce receive uniform application and coverage of waxes or oils using proper packing-line brushes or rolling sponges. Waxes applied to some fruits and fruit-like vegetables reduce water loss, replace natural waxes removed during washing, act as carriers for fungicides, minimize scuffing injury during sorting and grade handling, or improve the product's cosmetic appearance. Waxes must be approved food-grade materials. The two most common types are mineral oil–based waxes and natural waxes such as carnauba. They can be formulated as water-based emulsions or with a solvent. Solvent-based formulations dry quickly but their use may be restricted by air pollution regulations. Water emulsions are commonly used in California and require hot air drying after application. Studies indicate that if waxes reduce the rate of water loss by more than about one-third, they also reduce product gas exchange and may interfere with normal aerobic respiration, leading to off-flavors or decay. Formulations are available that have been approved for use in certified organic production, but in general these are only available in large, bulk containers, which may be problematic for smaller growers.

Packing and packaging can also be designed to minimize water loss. As discussed earlier, plastic bags are a popular way to improve unitized sales and help reduce water loss. To minimize condensation inside the bag and reduce the risk of microbial growth, the bags may be vented, microperforated, coated with a condensation dispersant, or made of material permeable to water vapor. However, it is important to remember that barriers to water loss may also function as barriers to cooling. Packing systems should be carefully selected for the specific application, with cooling and other postharvest handling needs in mind. The exposure of bagged or tightly wrapped produce to direct sunlight—or, to a lesser degree, to fluctuating temperatures during transport, distribution, and display—will cause the products' internal temperature to rise, potentially at a rapid rate. Water loss will result and, when cooling follows, free water will condense, generally leading to an unappealing growth of superficial mold or accelerated decay of the produce. During transportation and storage, relative humidity (or, more properly, vapor pressure deficit) is critical, even at low temperatures. For a more complete discussion of optimal relative humidity for fruits and vegetables and the principles for prevention of water loss, see *Commercial Cooling of Fruits, Vegetables, and Flowers* (ANR Publication 21567).

COOLING AND STORAGE

Without question, optimal temperature management is the single most important tool for maintaining postharvest quality. For products that are not field cured or exceptionally durable, the removal of field heat as rapidly as possible is the most highly desirable and optimal step that may be taken to maintain quality and prevent loss.

Harvesting cuts a commodity off from its source of water, but it is still alive and will lose water and turgor (firmness) through respiration and transpiration. Field heat can accelerate the rate of respiration, and with it, the rate of quality loss. Proper cooling protects quality and extends both the sensory (taste) and nutritional shelf life of produce. The capacity to cool and store produce gives the grower more flexibility both in harvest timing and in getting the product to market. When designing mechanical cooling facilities or scheduling for their use, growers and shippers have a tendency to underestimate the refrigeration capacity that will be needed to meet peak cooling demands. It is often critical that fresh produce rapidly reach the optimal pulp (internal tissue) temperature for short-term storage or shipping if it is to maintain its highest visual quality, flavor, texture, and nutritional content all the way to market. The four most common cooling methods suitable for small farms are described below.

Room cooling. For room cooling, growers use an insulated room or mobile container (such as a marine refrigerated shipping container) equipped with units designed to deliver conditioned air by controlled air flow. Room cooling is much slower than other methods. Depending on the commodity, packing unit, and stacking arrangement, the product may cool too slowly in some parts of the stack or packed unit to prevent water loss, premature ripening, or decay. By increasing air circulation in the room, using vented containers or boxes, and, where space allows, spacing the containers within the room or on pallets with vents aligned to allow airflow among stacked units, it is possible to increase the rate of cooling. As discussed earlier, any carton or bin liners or packaging materials have the potential to further extend the amount of time necessary to cool the product, and product in the middle of a container or pallet may never reach the desired temperature.

Forced-air cooling. Often called pressure cooling in some areas, forced-air cooling uses fans in conjunction with a cooling room to pull cold air through packages or bins of produce covered by a tarp to create a tunneling or air-channeling effect. Although the actual rate of cooling depends on the air temperature, the rate of airflow, package venting, and the volumetric size of the product, this method is usually 75 to 90 percent faster than simple room cooling. Design considerations for a variety of small- and large-scale units are available in *Commercial Cooling of Fruits, Vegetables, and Flowers* (ANR Publication 21567). Low-cost forced-air cooling systems are possible even for just a small stack of cartons, such as might be needed for cut flowers (see Kitinoja and Kader 2003).

Hydrocooling. The term *hydrocooling* refers to various system designs to shower produce with chilled water in order to remove heat and possibly clean the produce somewhat at the same time. The use of a disinfectant in the water is essential in recirculating systems; details of sanitizer use are discussed later in this chapter. Hydrocooling is not appropriate for all produce since some fruit and vegetables are susceptible to physical injury or water-soaking damage. Even for produce varieties that are not susceptible, the system must be carefully designed to prevent water-beating damage. Waterproof containers (such as plastic totes or RPCs) or water-resistant waxed corrugated cartons are required. Currently, waxed corrugated cartons have limited recycling or secondary use outlets, so reusable, collapsible plastic containers are gaining popularity. A list of products suitable for hydrocooling is available in *Postharvest Technology of Horticultural Crops* (ANR

Publication 3311) as well as in *Commercial Cooling of Fruits, Vegetables, and Flowers* (ANR Publication 21567).

Top icing or liquid icing. Icing is an effective method for cooling some commodities and is equally adaptable to small- and large-scale operations. Ice-tolerant produce varieties are also listed in *Postharvest Technology of Horticultural Crops* and in *Commercial Cooling of Fruits, Vegetables, and Flowers* and include broccoli, sweet corn, and a number of other vegetables. The ice you use must be free of chemical, physical, and biological hazards, and the packages and packing you use must be designed to suit this cooling method.

The criteria you need to consider when you select the appropriate cooling method and storage temperature and humidity conditions for perishable crops are, once again, discussed in *Postharvest Technology of Horticultural Crops* and in *Commercial Cooling of Fruits, Vegetables, and Flowers.*

IMPORTANCE OF OPTIMAL STORAGE AND SHIPPING TEMPERATURES

Although we stress rapid and adequate cooling and protection from heat gain as primary elements of postharvest handling, many fruits and vegetables produced in temperate-zone climates are subtropical in origin and susceptible to chilling injury. Chilling injury occurs when sensitive crops are exposed either before or after harvest to low temperatures (but above the freezing point) for an extended period. Even a brief exposure often causes damage, though it may not become apparent for several days or until the produce is transferred to warmer display conditions. Some examples of chilling-sensitive crops are basil, tomato, bell peppers, eggplant, green beans, honeydew melon, sweet potato, watermelon, okra, yellow crookneck squash, mango, cherimoya, and Fuyu persimmons. Individual parts of some commodities have distinct sensitivities. In eggplant, the cap or calyx is more sensitive and turns black before the fruit itself is affected. The effects of chilling injury are cumulative and become irreversible after exposure to a commodity-specific time-temperature combination. For fruits and fruitlike vegetables, the temperature and duration of exposure to cause chilling injury typically change with maturity and ripeness. Peaches, plums, and nectarines as well as honeydew melons and tomatoes, for example, show a great reduction in chilling defects as they increase in their degree of ripeness. Depending on the duration and severity of chilling, chilling symptoms become evident in the ways listed below a few hours or days after the product is returned to warmer temperatures. Symptoms may include some of the following:

- pitting and localized water loss
- browning or other skin blemishes
- internal water-soaking or discoloration
- increased susceptibility to superficial lesions from molds and decay
- failure to ripen or uneven color development
- loss of flavor, especially from characteristic volatiles
- development of off-flavors

POSTHARVEST ETHYLENE EFFECTS: POSITIVE AND NEGATIVE

Ethylene, a natural hormone produced by plants, is involved in many natural functions during preharvest growth, stress response, and induction of maturation and subsequent ripening. The management of ethylene is another postharvest consideration that can help you manage the timing of ripening events and maintain quality during storage and transportation. Ethylene treatments can be used to degreen citrus fruits and control ripening events in fruits harvested at a mature but unripe developmental stage, such as mature green tomatoes and pears. In organic handling, application of ethylene gas produced by catalytic generators has recently been allowed for tropical fruits such as bananas and mangoes, but not for subtropical fruits such as citrus and tomatoes. This has little relevance, however, for small-scale growers marketing locally and regionally, since the majority of their ethylene-responsive products are harvested nearly or fully ripe and have no need for ethylene application. For a detailed discussion of the role of ethylene in ripening and postharvest management, see *Postharvest Technology of Horticultural Crops* (ANR Publication 3311).

In contrast to the beneficial role of ethylene in controlling ripening, ethylene produced by harvested produce (either as a result of ripening events or of deterioration and decay) or introduced from environmental sources (e.g., exhaust from propane-powered lift trucks) can be very damaging to sensitive commodities. Ethylene-producing fruits should not be stored with fruits or vegetables that are susceptible to ethylene damage. Ethylene-sensitive commodities should not be handled in packinghouses or warehouses or held in cold storage areas where damaging concentrations of ethylene are present, whether due to combustion engine exhaust or venting of catalytic ethylene generators from degreening and ripening rooms. Ethylene concentrations of 1 ppm or more will stimulate loss of quality, reduced shelf life, and increased disease, and will induce specific symptoms of ethylene injury such as the following:

- russet spotting of lettuce
- yellowing or loss of green color (e.g., in cucumber, broccoli, kale, spinach)
- increased decay of radicchio
- increased toughness in turnips and asparagus spears
- bitterness in carrots and parsnips
- yellowing and abscission (dropping) of leaves in Brassicas (cabbage, many Asian greens)
- softening, pitting, and development of off-flavor in peppers, summer squash, and watermelons
- softening of avocado, Fuyu persimmons, and kiwifruit
- browning and discoloration in eggplant pulp and seed
- discoloration and off-flavor in sweet potatoes
- increased ripening and softening of fruits when not desired

Adequate venting or fresh air exchange is the most important way to minimize ethylene levels in a storage area. Another approach is to use an ethylene adsorption material or conversion system designed to prevent damaging levels (as low as 0.1 ppm for some commodities) from accumulating in storage rooms and transportation vehicles. Potassium permanganate ($KMnO_4$) is one material commonly used in air filtration for postharvest storage rooms or in blankets, pads, or individual sachets of pellets placed in cartons during transport. For greatest effectiveness, room or container air must be circulated through these filters. These $KMnO_4$ air filtration systems or sachet absorbers are allowed for organic postharvest handling, so long as handlers maintain strict separation (no actual contact) between the $KMnO_4$ materials and the organic product.

Other air filtration systems available for ethylene removal in cold rooms work by circulating room air through units that contain glass rods treated with a titanium dioxide catalyst and an ultraviolet light source that activates the catalyst and triggers ethylene destruction. Corona discharge or ultraviolet light–ozone-based purification systems are commercially available for both ethylene elimination and killing airborne spores. You can find sources of these materials in the online Postharvest Yellowpages (http://postharvest.ucdavis.edu).

SANITATION AND WATER DISINFECTION

Sanitation of equipment and food contact surfaces and disinfection of water used in processing should be integrated into every facet of postharvest handling in order to maintain quality, optimal storage life, and food safety. Adequate washing and cleaning of produce (for those fruits and vegetables that can tolerate postharvest water contact) to optimize customer presentation, control decay and spoilage, and minimize the risk of foodborne illness should be provided by growers or handlers at all scales of production. Preshipping washing is common with tolerant crops, especially in advance of any extended storage period or longer-distance distribution. Besides reducing decay pathogens, you can reduce the risk of (but not entirely eliminate) contamination by human pathogens including bacteria such as *Escherichia coli* (*E. coli*) O157:H7, *Salmonella, Shigella,* and *Listeria,* parasites such as *Cryptosporidium* and *Cyclospora,* and viral pathogens such as hepatitis virus A and norovirus by using a properly designed wash system in combination with approved antimicrobials. These and other pathogens have been isolated from or associated with illnesses that resulted from consumption of domestic and imported fresh vegetables. Cases of such illness are not unique to large-scale production and shipping or fresh-cut processors. We will discuss the topic of food safety in greater depth later in this chapter.

Harvest and postharvest water quality. The source, condition, and prior use of the water used for field trimming and packing operations such as hydration or harvest cut washing after field harvest or immediately before retail display, application of postharvest chemicals, and the use of washing and cooling water in packing sheds or value-added pack lines deserve special consideration and attention. The water used for all applications should be free of chemical, physical, and biological hazards. Assurance of its continuing high quality in any recirculation and reuse applications is essential to prevention of cross-contamination within a lot and among lots. Water quality is best maintained through integrated management of the cleanliness of incoming product, filtration, the addition of approved disinfection chemicals or treatments, the verification of dose by means of continuous or periodic measurement, and the periodic partial or complete clean-out and fresh water exchange of the system. A daily clean-out and sanitization of the system and complete water exchange is strongly recommended.

As a general practice, it is best to keep field soil on product, bins, totes, and pallets to a minimum by brushing or washing any adhering soil from harvest containers. Careful harvesting to minimize the presence of nonsalable plant material (vines, leaves, roots, damaged and decay product) will also help maintain water quality and reduce the need for more frequent flushing. Both steps are among the practices that will significantly reduce the demand (i.e., the amount of material that must be added for effective control or to achieve a target dose activity level) for disinfectant in the water and lower the total volume of antimicrobial agents required. A simple explanation and approach to hypochlorite (bleach) dose calculation for postharvest handling is described in *Water Disinfection: A Practical Approach to Calculating Dose Values for Preharvest and Postharvest Applications* (ANR Publication 7256). A light spray of water mixed with appropriate levels of needed disinfectants at harvest—for example, prewashing the butt end of lettuce or celery—removes plant exudates that have been released from harvest cuts or wounds, which can otherwise react rapidly with oxidizers such as hypochlorite and ozone, thereby neutralizing their effectiveness to some degree and requiring their reapplication at higher rates in order to maintain the originally intended effective level of activity.

Properly used, a disinfectant in postharvest wash and cooling water can help prevent both postharvest diseases and foodborne illnesses. Because most municipal water supplies are chlorinated and the vital role of water disinfection is well recognized, organic growers, shippers, and processors may use chlorine within specified limits. All forms of chlorinated materials (chlorine gas, liquid sodium hypochlorite, granular calcium hypochlorite, and chlorine dioxide) are restricted materials as defined by existing organic standards. Their application must conform to maximum residual disinfectant limit rules under the Safe Drinking Water Act, currently 4 mg/L (4 ppm), expressed as Cl_2. California Certified Organic Farmers (CCOF) regulations have in the past permitted this threshold of 4 ppm residual-free chlorine, measured downstream of the product wash. As this rule may change according to location and the degree of interpretation allowed, growers who are registered through the California Department of Food and Agriculture (CDFA) and certified by other agencies than CCOF should check with the state and their certifying agent. For a more complete discussion of water disinfection, see *Postharvest Chlorination: Basic Properties and Key Points for Effective Disinfection* (ANR Publication 8003) and *Making Sense of Rules Governing Chlorine Contact in Postharvest Handling of Organic Produce* (ANR Publication 8198).

Liquid sodium hypochlorite is the most common form of chlorine used in both organic and conventional operations. For optimum antimicrobial activity with a minimal concentration of applied hypochlorite, the pH of the water must be between 6.5 and 7.0, and you can add approved materials to adjust the pH into this range. At this pH range, there will be adequate amounts of chlorine in the form of hypochlorous acid (HOCl), which delivers the highest rate of microbial kill and minimizes the release of irritating and potentially hazardous chlorine gas (Cl_2). Chlorine gas will form and exceed safe levels if the water is highly acidic. Products used for pH adjustment include food-grade citric acid, hydrochloric acid (muriatic acid), phosphoric acid, sodium bicarbonate, lactic acid, and vinegar (acetic acid). Calcium hypochlorite, properly dissolved from granular or tablet form, may provide the benefit of reducing sodium injury to sensitive crops (e.g., some apple varieties, light-skinned cucumbers, immature yellow squash), and limited evidence points toward reduced decay from wound-healing and extended shelf life for tomatoes and bell peppers as a result of calcium uptake from the application. Useful guides, tables, and links to other sites that can help you determine the amounts of sodium or calcium hypochlorite to add to clear, clean water for disinfection per makeup water and total use volume are available from several sources, including the following:

UC Postharvest Technology, Research, and Information Center. http://postharvest.ucdavis.edu

UC Vegetable Research and Information Center. http://vric.ucdavis.edu/veginfo/veginfor.htm

ATTRA National Sustainable Agriculture Information Service. http://attra.ncat.org

Oxidation-reduction potential. Another method for monitoring, controlling, and documenting the chlorine dose and pH status of any water is oxidation-reduction potential (ORP), or the redox potential, measured in millivolts (mV). Some operators specify both a redox potential target (≥650 mV) and a pH window of operation (pH 6.5 to 7.5), but the latter is an inaccurate measurement in the case: the mV value is the determining criterion. Sensors for ORP measure the oxidizing (or reducing) potential or activity of a solution. The higher a water's ORP value, the greater the oxidizing action and the shorter the microbial kill time in water. The relationship between ORP and traditional dose measurements (parts per million [ppm]) is nonlinear in the range most commonly used in postharvest management. This means that you typically cannot use ORP millivolt values to demonstrate that you have met a set ppm specification. For a more complete discussion of ORP benefits and limitations, consult *Oxidation-Reduction Potential (ORP) for Water Disinfection Monitoring, Control, and Documentation* (ANR Publication 8149).

Oxidation-reduction potential has been found to be a functional and practical single-value measurement of disinfection power under most conditions. It can be a verifiable, successful method for assessing the antibacterial and antifungal status of washout, flume, washline, or cooling water systems. As mentioned earlier, ORP measurements in surveys of commercial systems have not proven to provide a direct or repeatable correlation to an exact concentration of free chlorine, regardless of water pH. Tests do, however, demonstrate that the ORP status of water over a broad pH range correlates well with microbial lethality or control that results from the oxidizing activity of the solution. Clearly, ORP sensors, like any equipment, need to be calibrated, maintained, and replaced as necessary if you want accurate readings. ORP sensors, particularly in-line probes, may be prone to oxidizer saturation or fouling and may therefore give readings that do not reflect the current water quality conditions. It is always advisable—if not essential—to have redundant systems for verification that the process is under control. A well-managed system includes periodic testing with hand-held ORP and pH sensors at multiple points along with cross-checks against direct measurements of available free chlorine to make sure all measurements agree within a practical range.

Rapid pathogen inactivation will occur at a specific threshold ORP, whether the water pH status is above or below 7.0, and this is well documented in the literature and matches the chapter authors' applied research experience. Among hand-held and in-line device results, the ORP value is the best measure of the availability of the hypochlorous acid (HOCl) form of chlorine in water at any given pH. HOCl is reported to be at least 80 times stronger at killing harmful microbes than the hypochlorite ion (OCl^-) form, which is more abundant at elevated pH levels (approximately 20% HOCl and 80% OCL^- at pH 8.0).

The accuracy of ORP for detecting levels of surviving microbes is not perfect or without occasional uncertainties in total microbial inactivation rates when you move from single-pass water to multiple-use recirculated water, the latter of which accumulates suspended organic and inorganic solids and salts. This lack of accuracy is particularly evident in water with a high sediment content and short residence (contact) time prior to chlorine neutralization and after sample collection. You can get a reasonably short response time and be free of any wide flux or instability in the measured values from ORP sensors for a moderate range of water quality consistent with the conditions in many packing and wash process systems. However, in surveys of packing sheds with larger recirculating systems, the authors have found that the frequently cited target of 650 mV is often too hard to maintain uniformly and is too low a value to adequately meet bacterial and fungal inactivation objectives. Depending on the specific application, product, and microbial control objectives, users often set the injection point dose targets or set points at 725 to 850 mV. With increasing water complexity (i.e., poorer-quality water), the concentration of free chlorine necessary to maintain a constant or target ORP increases.

Nonchlorine oxidizers for water disinfection. Ozone is an attractive option for water disinfection and other postharvest applications, and you can find a detailed discussion of postharvest ozone in *Ozone Applications for Postharvest Disinfection of Edible Horticultural Crops* (ANR Publication 8133). Ozonation, a powerful oxidizing treatment, is effective against chlorine-tolerant decay microbes such as some *Fusarium* and *Alternaria* spore forms and against foodborne pathogens, and acts quickly in clean water systems. The reactivity of ozone may give it a distinct advantage in cooling or wash procedures that feature a short contact time or in long

RECOMMENDED STEPS TO OPTIMIZE POSTHARVEST CHLORINATION*

1. Minimize all sources of chlorine demand (soil, plant debris, heavily wounded or decayed produce) on incoming product.
2. Inspect incoming product during precooling stage for excessive amounts of adhering soil and nonproduct plant material (such as leaf or vine trash) on totes, cartons, and pallets. Remove as practical.
3. Provide feedback as needed to supervisors of harvest operations and crews to improve performance in reducing sources of chlorine demand at the field level.
4. For fruit and vegetable products that will tolerate it, light mechanical cleaning (such as dry brushing or brush washing) may significantly reduce chlorine demand and extend the clarity of the process water and the functional disinfection activity.
5. Manage and monitor postharvest water to maintain a pH of 6.5 to 7.0 and a level of free chlorine sufficient to achieve disinfection goals. Organic handlers should not exceed 4 ppm free chlorine, measured at the point where the product is removed from that operational step. For example, adjust the chlorine level in dump and flume tank injection water to maintain 50 ppm and maintain the chlorine level in recirculating water at 4 ppm or less at the return sump (the farthest point from injection—for example, where the product transfers to a lift conveyor).
6. For products that will tolerate it (such as tomatoes and melons), heated chlorinated water is more effective for disinfection. This benefit must be balanced, however, against reduced stability of the chlorine and increases in irritating chlorine levels from off-gassing. Heating the initial receiving water to 10°F above the temperature of the incoming product also minimizes the potential for water infiltration into the product, which may trigger or accelerate decay and presents a known risk for foodborne illness if pathogens are present.
7. If necessary, add approved flocculants to capture suspended sediments in the water and hold the water in a retention sump or basin. Screening and filtration of recirculating water will reduce chlorine demand and improve the performance of any oxidizer (such as ozone and peroxides) or nonoxidizing disinfectant (such as a UV system).
8. Dump tank, flume, and hydrocooler sump basins and any sediment collection points in the equipment must be cleaned daily. Sediments are a common reservoir for decay pathogens and pathogens of concern for human food safety.
9. Develop a system of periodic partial or full replacement with clean water that balances the costs and time delays involved in cooling or heating postharvest water, with the goal of minimizing turbidity and electrical conductivity (salt buildup) in the water. Develop a simple rating system to assess turbidity thresholds for periodic water exchange. A simple turbidity tube with standard Secchi disk (with a black-and-white pattern viewed through the tube) is very affordable and easy to make. (See http://www.cee.mtu.edu/sustainable_engineering/resources/technical/Turbidity-Myre_Shaw.pdf.)
10. Ensure that responsible personnel are trained in the function and operation of equipment and monitoring kits. Define the timing and frequency of procedures to be used in monitoring in written form and post standard procedures for each water use, whether it involves using a bucket in the field, a small wash line, or a more extensive chopping, slicing, and packaging system.
11. Use redundant systems of measurement to ensure adequate dosing, such as calibrated ORP sensors and free-chlorine and pH test strips or titration kits.
12. Ensure that responsible personnel are trained to safely handle concentrated chemical sanitizers and to implement and document any corrective actions that must be taken to meet product quality and safety.

*For a discussion of issues related to chlorine use and organic integrity, see *Making Sense of Rules Governing Chlorine Contact in Postharvest Handling of Organic Produce* (ANR Publication 8198).

flume systems. Organic postharvest handlers may find they prefer to use ozonation rather than chlorination since it may help them meet the requirements of certain markets that do not allow food that has been in contact with chlorine or help them appeal to particular consumer sectors. Ozone oxidative reactions create far fewer disinfection by-products than chlorination (e.g., trihalomethanes, which are a health and environmental concern). However, be aware that capital and operating costs are typically higher for ozonation than for chlorination or other available methods that do not require on-site equipment.

Ozone must be generated on-site at the time of use and is very unstable, lasting as few as 20 minutes even in clear water, and so has no residual activity. This rapid breakdown to molecular oxygen (O_2) is desirable because it means the ozone is nonpolluting, but naturally it also means that maintaining effective ozone levels even in small volumes of water is very challenging. Still, it can be effective. Ozone systems are in use by fruit and vegetable packers and fresh-cut processors, both conventional and organic, typically at a focused point in the handling process. Ozonation of the final rinse water has become fairly common. Clear water is essential for optimal performance, and the process requires adequate to superior filtration of input or recirculating water. Depending on scale and ozone generation output, costs for a complete system start at about $10,000. Small-scale units available for a few thousand dollars are suitable for limited water use and small-batch applications. For specifications and installation, consult an experienced ozone service provider. Sources of these services and suppliers can be found in the online Postharvest Yellowpages (http://postharvest.ucdavis.edu).

Food-grade hydrogen peroxide (0.5 to 1%) and peroxyacetic acid are additional options for nonchlorine treatment. In general, peroxyacetic acid (PAA; 11% hydrogen peroxide, 15% acetic acid) has good efficacy in water dump tanks and water flume sanitation applications. Reports indicate that PAA has very good performance, when compared to chlorine and ozone, in killing yeasts and molds and in removing and controlling microbial biofilms (tightly adhering slime) in dump tanks and flumes. These peroxide formulations have a higher per-unit cost than hypochlorite. Approved peroxide-based materials are available for organic uses, and as a result of growing interest in nonchlorine alternatives, treatment containers are now available in volumes designed for use in smaller operations.

Other examples of organically allowed postharvest treatments include organic acids from natural sources (e.g., acetic acid–vinegar, malic acid), spice extracts and plant essential oils (e.g., rosemary, thymol, clove, spearmint, peppermint, cinnamaldehyde), thiosulfinates (e.g., allicin and garlic extracts), and copper ions. Commercial formulations that include these antimicrobials are available.

Planning for Postharvest Safety

Planning for postharvest food safety should be included in any edible crop management plan. Regardless of the scale of operations or whether conventional, organic, biodynamic, or any other crop management approach is practiced, no farming operation should be viewed as exempt or excluded from the regulatory requirements, consumer expectations, and moral responsibilities that bear on food safety. Beyond the commonsense notions that we all learn regarding health, hygiene, and good crop husbandry, specific prerequisite programs that are consistent with good agricultural practices (GAP) need to be understood, customized, and formalized by each grower, handler, and operator and for each crop and specific production field. This will help minimize the risks associated with a variety of hazards and contaminants, including chemicals (e.g., heavy metals carryover), physical contaminants (e.g., sand and soil, wood, glass, plastic or metal shards), and biological hazards (e.g., pathogenic *E. coli, Salmonella, Listeria,* parasites, mycotoxins). You can find a detailed outline of the key points and principles of preharvest and postharvest GAPs in *Key Points of Control and Management of Microbial Food Safety: Information for Growers, Packers, and Handlers of Fresh-Consumed Horticultural Products* (ANR Publication 8102) and Suslow et al. (2003), as well as other resources available from the National GAPs Program Web Links at Cornell University (http://www.gaps.cornell.edu/gapsd/Weblinks.html) and UC Davis (http://ucgaps.ucdavis.edu).

In addition to noting that many elements of a GAP plan are likely to be incorporated into their existing crop management program and activities, growers commonly realize improvements in efficiency and benefits to product quality after they implement a GAP and food safety system. Programs already in place to ensure produce quality may, with minor modifications, be made to apply directly to food safety. Though it has rarely been documented, growers often share strong empirical evidence indicating that the application of food safety programs has proven to have a direct benefit on postharvest quality and market access.

Chemical safety. Chemical inputs may be needed to fertilize crops or control pests during production and postharvest handling. These products must always be used according to product labeling to avoid unsafe levels of residues. The label will tell you which products the chemical may be applied to, how close to harvest the materials can be applied, and the proper application rate. Take particular care when using several products at the same time. Take preventive measures to ensure that mechanical lubricants, fuels, and solvents used around the farm and on equipment do not come into contact with produce handling surfaces or into direct contact with product in the field, at harvest, during packing, or during transportation.

Physical safety. Harvesting and packaging processes should be managed to keep harvested products from becoming contaminated with foreign matter such as plastics, wood chips, rocks, and glass or metal fragments.

Biological safety. Prior land use, adjacent land use, water source and method of application, fertilizer choice (such as the use of manure), compost management, equipment maintenance, field sanitation, movement of workers between different operations, personal hygiene, domestic animal and wildlife activities, and other factors all have the potential to adversely impact food safety. A variety of resources available to growers are listed later in this chapter under "Other Resources," and you can use them to guide you in the development of a GAP program and on-farm self-audit. All growers, handlers, and marketers of fresh vegetables should be aware of the U.S. Food and Drug Administration (FDA) Primary guide to GAPs, which is available in English, Spanish, Hmong, Lao, Llocano, Portuguese, French, and Arabic (Portal to FDA Food Safety Information, http://www.fda.gov/food/foodsafety).

Packing and cold storage facility operators need to make special considerations for management of *Listeria monocytogenes,* due to its common presence in the field on vegetation, in soil, and in cool, wet facilities and drains, and its ability to grow on surfaces, equipment, and most fresh produce under refrigeration temperatures *Guidelines for Controlling* Listeria monocytogenes *in Small- to Medium-Scale Packing and Fresh-Cut Operations.* (See ANR Publication 8015).

SUMMARY

To protect your investment in the fruits and vegetables you have produced, you need to ensure that they receive careful treatment after harvest. Handle produce in a gentle but expedited fashion in order to both reduce the time from harvest to cooling and prevent physical damage. Optimum postharvest quality and safety begins with the selection of the appropriate varieties and the use of good agricultural practices. Good temperature management and a shortened time from harvest to market are key factors to maintaining fruit and vegetable appearance, flavor, and nutritional quality.

FOR SPECIALTY CROPS, MAKE EDUCATED GUESSES

When dealing with new crops and determining how they should be handled postharvest, one can make a few educated guesses based on the answers to the following questions:

1. Is the crop of tropical or temperate origin? This will likely tell you whether it is chilling sensitive.
2. Is the crop a leaf, a root, or a fruit? This can help determine how susceptible it is to water loss.
3. If the crop is a fruit, are there noticeable ripening changes after harvest? The degree of change after harvest is generally related to its rate of deterioration.
4. Are you harvesting the crop when it is rapidly growing or when it has completed its growth phase? Rapidly growing crops generally have a very high respiration rate and high deterioration rate.
5. If the crop is a leafy product, are there rapid color changes? This may indicate how susceptible the product is to deterioration and how sensitive it may be to exposure to the contaminant ethylene.
6. If the crop is a fruit, are there rapid textural and compositional (starch-to-sugar conversion) changes? This may indicate a climacteric type fruit that produces a lot of ethylene.
7. What are the postharvest characteristics of a related product (another species of the same genus, another genus of the same family, etc.)? Refer to tables 10.1 and 10.2 for information on various products.
8. What is the estimated storage temperature? Try to place the product into one of the following categories:

   A. low temperature (32° to 41°F)

   B. moderate temperature (41° to 50°F)

   C. moderately high temperature (50° to 60°F)
9. What is the estimated shelf life? Try to fit it into one of the following categories:

   A. short shelf life: 1 to 6 days

   B. moderate shelf life: 7 to 21 days

   C. long shelf life: 3 to 12 weeks or longer
10. Is the product very tender and delicate? Does it bruise easily? This will help to determine what sort of packaging system might be appropriate.

Examples of Defects that Do Not Affect Postharvest Life Potential of Fresh Produce

- healed frost damage
- healed scars and scabs
- well-healed insect stings
- irregular shape
- suboptimal color uniformity or intensity

Examples of Defects that Do Affect Postharvest Life Potential of Fresh Produce

- softening
- sunburn and sunscald
- cracks
- bruising
- sprouting
- chilling injury
- scald
- decay
- cuts, abrasions, and skin breaks

COMPARISON OF POTENTIAL ADVANTAGES AND DISADVANTAGES OF PLASTIC CONTAINERS VS. CARTON BOXES

Potential advantages	Potential disadvantages
PLASTIC CONTAINERS	
Significant cost reduction in closed loop systems	Less product protection and cushioning; need for packing materials and interior materials
Returnable, reusable, and recyclable	High initial investment
Easy to clean and sanitize	Increased shipping costs due to weight and back-shipping
Any size and color; color-code products	Mandatory cleaning
Moistureproof; no collapsing or resulting product damage	Tracking and inventory costs; storage space requirements
Reduced labor costs and handling steps	Replacement costs due to theft, damage
CARTON BOXES	
Product protection good in a well-designed carton box	Stacking strength greatly reduced under high moisture conditions
Can print with high-resolution graphics for attractive product merchandising	Handling costs at distribution centers to recycle
Can custom design containers rapidly	Box-making labor and equipment costs
Low shipping costs; no back-shipping cost	Not returnable and reusable
Storage of cartons is space efficient	Wax-impregnated and other carton containers difficult to recycle
One-way container, no cleaning	Cannot be cleaned

Source: Compiled by Marita Cantwell from various industry sources, including Fiber Box Association (http://www.fibrebox.org).

Table 10.1. Examples of postharvest requirements for selected vegetables and melons

Product	Harvest quality	Storage		Shelf life (days)	Ethylene sensitivity	Observations
		°F	% RH			
Artichoke, globe	size, tender bracts	32	95	14	low	sprinkle lightly
Asparagus	bracts at tip closed	36	95	14	low	stand in water
Basil	fresh, tender leaves	55	95	7	moderate	stand in water
Beans, lima	seeds developed, plump	40	95	7	moderate	sprinkle lightly
Beans, pole and snap	crisp pods, seeds immature	40	95	7	moderate	sprinkle lightly
Beets, bunched	firm, deep red roots	32	95	14	low	sprinkle, cut tops
Broccoli	firm head, buds not open	32	95	14–21	high	sprinkle; ice
Brussels sprouts	firm sprouts	32	95	21–28	high	sprinkle; ice
Cabbage	crisp, firm, compact head	32	95	30–180	high	sprinkle lightly
Cantaloupe melon	stem separates; rind color	36	95	14	moderate	ice
Carrots, topped	tender, crisp, sweet roots	32	95	28–180	high	sprinkle; cut tops
Cauliflower	compact, white curds	32	95	14–21	high	sprinkle
Celery	crisp, tender petioles	32	95	14–21	moderate	sprinkle; ice
Corn, sweet	plump tender kernels	32	95	7	low	ice
Cucumber	crisp, green, firm	50	95	10	high	sprinkle lightly
Eggplant	seeds immature; shiny, firm	50	95	10	moderate	
Endive, escarole	fresh, crisp, tender leaves	32	95	14–21	moderate	sprinkle lightly
Greens, leafy, and herbs	fresh, crisp, tender leaves	32	95	10–14	moderate	sprinkle lightly
Honeydew melon	waxy, creamy colored, heavy	45	90	21	high	
Lettuce	compact head, crisp, tender	32	95	21	high	sprinkle lightly
Onions, dry	firm bulbs, tight necks	32	65	30–180	low	
Onions, green	crisp stalks, firm white bulbs	32	95	10	moderate	sprinkle; ice
Parsley	crisp, dark green leaves	32	95	21	high	sprinkle; ice
Peas	tender, green, sweet pods	32	95	7–10	moderate	sprinkle
Peppers, chili	firm with shiny appearance	45	95	14	low	
Peppers, green	firm with shiny appearance	45	95	14	low	
Potatoes, early crop	well-shaped tubers, defect-free	50	90	14	Low	if washed, dry well
Potatoes, late crop	well-shaped tubers, defect-free	45	90	60–180	moderate	if washed, dry well
Pumpkin	hard rind, good color, heavy	55	65	30–160	moderate	
Radish, with tops	firm, crisp, dark green leaves	32	95	14–21	Moderate	sprinkle; ice
Rutabagas	roots firm with smooth surface	32	95	60–120	low	cut tops; sprinkle
Spinach	dark green, fresh, crisp leaves	32	95	10	high	sprinkle lightly
Squash, summer	firm, shiny fruits, right size	45	95	10	moderate	
Squash, winter	hard rind, corked stem, heavy	55	65	60–120	moderate	allow cut stems to heal
Tomatoes, green	firm, jelly present, light green	55	90	21	high	
Tomatoes, ripening	firm, uniform coloration	50	90	14	high	avoid temperatures <50°F
Turnip	firm, heavy roots	32	95	60–120	low	cut tops; sprinkle
Watermelon	crisp, good flesh color, heavy	55	90	14	high	

Table 10.2. Examples of postharvest requirements for selected fruits

Product	Harvest quality	Storage °F	Storage % RH	Shelf life (days)	Ethylene sensitivity	Observations
Apple	crisp, color typical of variety	32	95	90–180	high	varieties differ a lot in postharvest life
Apricot	firm, well colored	32	95	7–21	moderate	
Avocado	% dry matter, % oil, size	41	90	14–28	high	varieties differ a lot in chilling sensitivity
Banana	finger size, color	55	95	7–14	moderate	
Blueberry	blue color, firmness	32	95	14–21	low	
Carambola	yellow skin color	40	95	14–21	moderate	
Cherimoya	firmness, skin color	55	95	14–21	high	
Cherry, sweet	fruit color typical of variety	32	95	7–14	low	
Cranberry	fruit color	36	95	30–60	low	
Currant, gooseberry	firm, color typical of variety	32	95	10–21	low	
Date	color, sugar content	32	75	180–360	low	
Fig	firm but near ripe, skin color	32	95	7	low	
Grape, table	color typical of variety, sugar	32	95	14–90	low	gray mold ends life
Grapefruit	skin color, sugars-to-acid ratio	55	95	30–40	low	
Guava	skin color, firm	45	95	14–21	moderate	varieties differ a lot
Kiwifruit	firmness, soluble solids	32	95	120–150	high	
Lemon	juice content	45	95	30–120	moderate	
Lime	juice content, skin color	50	90	21–50	high	
Litchi	red skin color, soluble solids	36	95	14–28	moderate	
Loquat	fruit color typical of variety	32	95	14–28	low	
Mandarin	peel color, sugars-to-acid ratio	41	95	14–28	low	
Mango	skin color, shape for variety	55	95	14–21	moderate	
Nectarine	skin ground color, firmness	32	95	14–28	moderate	
Olives	color (green or black)	45	95	14–28	low	harvested olives are processed
Orange	skin color, sugars-to-acid ratio	41	95	30–60	moderate	
Papaya	skin color	55	95	14–21	High	
Passion fruit	skin color change	45	90	7–14	moderate	
Peach	skin ground color	32	95	14–28	moderate	
Pear, Asian	firm, skin color of variety	32	95	30–120	moderate	
Pear, European	firmness, skin color	32	95	30–120	High	
Persimmon	skin color	32	95	14–28	High	
Pineapple	skin yellowing; "eye" flatness	50	95	7–21	moderate	
Plum, fresh prune	skin color typical of variety	32	95	10–30	moderate	
Pomegranate	size, skin color	45	95	60–120	low	
Prickly pear cactus	firm, jelly present, light green	41	95	14–21	low	
Quince	skin ground color changes	32	90	30–90	low	
Strawberry	fruit color	32	95	7–10	low	
Tamarillo	peel and pulp color	41	90	14–28	low	
Tamarind	pulp and shell brown, brittle	68	75	14–28	none	keep 6 months at 41°F

RESOURCES AND REFERENCES

Postharvest Practices

Bachmann, J., and R. Earles. 2000. Postharvest handling of fruits and vegetables. ATTRA National Sustainable Agriculture Information Service. Pp. 19. http://attra.ncat.org/attra-pub/PDF/postharvest.pdf.

Gross, K. C., C. Y. Wang, and M. Saltveit, eds. 2004. Commercial storage of fruits, vegetables, and florist and nursery stocks. USDA Agricultural Handbook #66. http://www.ba.ars.usda.gov/hb66.

Hoppe, R. A., P. Korb, E. J. O'Donoghue, and D. E. Banker. 2007. Structure and finances of U.S. farms: Family farm report, 2007 edition. USDA–ERS, Economic Information Bulletin No. (EIB-24). Pp. 58.

Kader, A. A., tech. ed. 2002. Postharvest technology of horticultural crops. Third edition. University of California Division of Agriculture and Natural Resources, Oakland. Publication 3311. Pp. 535. http://anrcatalog.ucdavis.edu/Postharvest/3311.aspx.

Kitinoja, L., and A. A. Kader. 2003. Small-scale postharvest practices: A manual for horticultural crops. Fourth edition. University of California Postharvest Technology Research and Information, Center, Publication No. 8E (also available in Spanish, Vietnamese, Khmer, Chinese, Punjabi, and Arabic). Pp. 267. http://postharvest.ucdavis.edu/Pubs.

Miles, A., and M. Brown, eds. 2005. Teaching direct marketing and small farm viability: Resources for instructors. Center for Agroecology and Sustainable Food Systems, UC Santa Cruz. Pp. 312. http://casfs.ucsc.edu/education/instructional-resources/teaching-direct-marketing-and-small-farm-viability.

Suslow, T. 1997. Postharvest chlorination: Basic properties and key points for effective disinfection. University of California Division of Agriculture and Natural Resources, Oakland. Publication 8003. Pp. 8. http://anrcatalog.ucdavis.edu.

———. 2000. Postharvest handling for organic crops. University of California Division of Agriculture and Natural Resources, Oakland. Publication 7254. Pp. 8. http://anrcatalog.ucdavis.edu/pdf/7254.pdf.

———. 2001. Water disinfection: A practical approach to calculating dose values for preharvest and postharvest applications. University of California Division of Agriculture and Natural Resources, Oakland. Publication 7256. Pp. 4. http://anrcatalog.ucdavis.edu/pdf/7256.pdf.

———. 2004. Oxidation-reduction potential (ORP) for water disinfection monitoring, control, and documentation. University of California Division of Agriculture and Natural Resources, Oakland. Publication 8149. Pp. 5. http://anrcatalog.ucdavis.edu/Items/8149.aspx.

———. 2004. Ozone applications for postharvest disinfectant edible horticultural crops. University of California Division of Agriculture and Natural Resources, Oakland. Publication 8133. Pp. 8. http://anrcatalog.ucdavis.edu/pdf/8133.pdf.

———. 2006. Making sense of rules governing chlorine contact in postharvest handling of organic produce. University of California Division of Agriculture and Natural Resources, Oakland. Publication 8198. Pp. 6. http://anrcatalog.ucdavis.edu/pdf/8198.pdf.

Thompson, J. F., P. E. Brecht, and T. Hinsch. 2002. Refrigerated trailer transport of perishable products. University of California Division of Agriculture and Natural Resources, Oakland. Publication 21614. Pp. 28. http://anrcatalog.ucdavis.edu.

Thompson, J. F., P. E. Brecht, T. Hinsch, and A. A. Kader. 2000. Marine container transport of chilled perishable produce. University of California Division of Agriculture and Natural Resources, Oakland. Publication 21595. Pp. 32. http://anrcatalog.ucdavis.edu.

Thompson, J. F., F. G. Mitchell, T. R. Rumsey, R. F. Kasmire, and C. H. Crisosto. 2001. Commercial cooling of fruits, vegetables, and flowers. University of California, Division of Agriculture and Natural Resources, Oakland, CA. Publication 21567. Pp. 61. http://anrcatalog.ucdavis.edu

Thompson, J. F., and M. Spinoglio. 1996. Small-scale cold rooms for perishable commodities. University of California Division of Agriculture and Natural Resources, Oakland. Publication 21449. Pp. 9. http://anrcatalog.ucdavis.edu.

Tourte, L., M. Gaskell, R. Smith, C. Fouche, S. T. Koike, and J. Mitchell. 2006. Organic certification, farm production planning, and marketing, University of California Division of Agriculture and Natural Resources, Oakland. Publication 7247. Pp. 5. http://anrcatalog.ucdavis.edu/pdf/7247.pdf.

Food Safety and Good Agricultural Practices

National Good Agricultural Practices Program (GAPS). http://www.gaps.cornell.edu/gapsd/Weblinks.html.

Suslow, T. 2003. Key points of control and management of microbial food safety: Information for growers, packers, and handlers of fresh-consumed horticultural products. University of California Division of Agriculture and Natural Resources, Oakland. Publication 8102. Pp. 7. http://anrcatalog.ucdavis.edu/pdf/8102.pdf.

Suslow, T., and L. Harris. 2000. Guidelines for controlling Listeria monocytogenes in small- to medium-scale packing and fresh-cut operations. University of California Division of Agriculture and Natural Resources, Oakland. Publication 8015. Pp. 8. http://anrcatalog.ucdavis.edu/pdf/8015.pdf.

Suslow, T., M. P. Oria, L. R. Beuchat, E. H. Garrett, M. E. Parish, L. J. Harris, J. N. Farber, F. F. Busta. 2003. Production practices as risk factors in microbial food safety of fresh and fresh-cut produce. Comprehensive Reviews in Food Science and Food Safety 2S:38–77.

USDA ERS. 2007. USDA Good agricultural practices and good handling practices audit verification checklist. http://www.ams.usda.gov/AMSv1.0/getfile?dDocName=STELPRDC5050869.

U.S. FDA. FDA/DFSAN Produce Food Safety Guidance Documents. http://dfa.gov/food/foodsafety/Product-SpecificInformation/FruitsVegetablesJuices.

U.S. FDA. Portal to FDA Food Safety Information. http://www.fda.gov/food/foodsafety.

University of California Good Agricultural Practices (UC GAP). http://ucgaps.ucdavis.edu.

Other Resources

ATTRA National Sustainable Agriculture Information Service. http://attra.ncat.org.

California Certified Organic Farmers, Certification Handbook. http://www.ccof.org/certification.php.

Organic Materials Review Institute. http://www.omri.org.

UC Postharvest Technology Research and Information Center. http://postharvest.ucdavis.edu.

11
Raising Animals

JAMES OLTJEN, FRANCINE BRADLEY, ROGER INGRAM, HOLLY GEORGE, BARBARA REED, CAROL COLLAR, DEBORAH GIRAUD, JOHN HARPER, ERIC MUSSEN, JOSH DAVY, AND DEANNE MEYER

Chickens, pigs, rabbits, goats, sheep, and sometimes horses and cows, are typical animals on a small farm. These animals complement the small farm; they eat crop residues or feed grown on-farm, and their waste, when mixed into the soil, recycles nutrients for plant use and serves as organic matter to improve soil structure. Your livestock enterprise can provide food for your family, supplement income, and provide diversity to your operation. The animals may also be pets, provide recreation, or you can use them to manage vegetation for weed or fire control. The choice of the right animals—those that are best suited to your needs and abilities—will help determine whether you will benefit or make a profit. Talk to others who are already raising the livestock you are considering. Listen to their histories and learn from them.

As a small farmer with limited land, facilities, and finances, you may want to consider developing a market for livestock products that are not widely available, such as locally or organically produced meat or eggs, free-range chickens, or miniature horses. Information may not be easy to find, but your University of California Cooperative Extension Livestock Farm Advisor can give you information about similar breeds which will help you get started.

BASIC CONSIDERATIONS

Legal restrictions. Check state and local restrictions regarding the types and numbers of animals that may be kept on your property. There may be restrictions on animals' proximity to wells and dwellings, and there are often rules on fences and structures for animals.

Other legal problems may come up after you establish a livestock operation. Neighboring dogs or resident wildlife can sometimes harass or kill your animals. Sometimes neighbors complain about noise, dust, odor, or flies. If you fail to manage these conditions correctly, they can lead to legal action.

The best way to familiarize yourself with local rules is to start collecting information from regulatory agencies. Ask your local county UC Cooperative Extension Livestock Farm Advisor for suggestions or contacts. Also contact county or city departments of planning and zoning and the county agricultural commissioner to see if your property has animal restrictions. Check with state offices, including the state brand inspector at the Department of Food and Agriculture for cattle, and the Department of Fish and Game, to learn of applicable regulations. These agencies are listed in the government section of your telephone book and can be found on the Internet. When you call, describe your situation and ask how your proposed operation might be affected by their regulations. Ask if they know of other departments you should contact.

Purchase price for animals, and where to buy. The purchase price of an animal varies with species, stage of maturity, health and condition, geographic location, sex, and whether it is a commercial, a purebred or a registered animal. Commercial animals are animals with parents of no particular breed. Purebred animals have parents of the same breed and often are registered with a breed association. Registered animals usually cost more than unregistered (or grade) animals because of the cost of maintaining specific traits of a particular breed and the market value of those traits.

Generally, the smaller the animal, the lower the price—you wouldn't expect to pay more for a rabbit than for a calf. Younger animals and very old animals are sometimes less expensive because their risk of mortality is greater. Animals in poor health and condition are cheaper, but generally are not recommended. Geographic location can influence the price of animals because of local supply and demand conditions.

Local or state livestock associations can provide you with the names of people who raise livestock for resale. Registered breed associations will have a membership list and will give you the names of breeders in your area.

Feed stores usually have a public bulletin board with advertisements of livestock for sale. Auction yards have sales at least weekly. UC Cooperative Extension Farm Advisors with knowledge of livestock will know about many of these and can provide you with advice on local sources.

Feed costs. Feed is usually the greatest variable expense a livestock owner will incur. Your decision about what kind of animal to raise will be based partially on the volume of feed the animal will consume and your financial ability to provide it. Expect to pay even more if you are producing organic animal products, since state and federal laws require the use of organic feedstuffs, which are generally more expensive. The amount of food needed is related to the animal's body size. Nutrient requirements (particularly for energy and protein) vary depending on the animal's stage of life. Different amounts of nutrients are required if the animal is immature, pregnant, or lactating. The larger the animal, the larger its intake and the higher the cost to feed it.

Ruminant animals—those with multicompartmented stomachs, like goats, sheep, and cattle—and near-ruminants, like rabbits or horses, can use low-quality feed and need only one or two types of feed. Fowl and hogs have more complex dietary requirements and need more complete rations. These higher-quality diets cost more.

Using pasture or range. If you want your livestock to graze, you need to determine how much pasture or range area the animals will need. The amount of land varies depending on species and life stage as well as whether the land is irrigated, its soil type and topography, the local climate, and land and animal management practices. To determine land requirements for different types of animals, use the concept of animal unit equivalents (AUE) (table 11.1).

Animal terminology. The following are terms that are helpful to know when raising animals:

- **Animal unit (AU).** The equivalent of one mature (1,000-lb) cow that consumes an average of 2.5 percent of its body weight (i.e., 25 lb) of dry feed per day.

- **Animal unit month (AUM).** The potential air-dried forage intake of 1 animal unit (1,000-lb animal) for 30 days, or about 750 pounds (2.5% x 1,000 lb x 30 days = 750 lb).

- **Stocking rate.** The amount of land allocated to each animal unit for the entire grazeable period of the year.

Table 11.1. Approximate daily forage consumption, by animal, given in animal unit equivalents (AUE)

Animal	Daily consumption (AUE)
Cows	
Bull, 24 month, 1,700 lb	1.50
Bull, 18-24 month, 1,150 lb	1.15
Cow and calf pair	1.35
Cow, mature, nonlactating, 1,000 lb	1.00
Heifer, 18 month, pregnant, nonlactating	1.00
Yearling, 18-24 month, 875 lb	0.90
Yearling, 15-18 month, 750 lb	0.80
Yearling, 12-15 month, 625 lb	0.70
Calf, weaning to 12 month, 500 lb	0.60
Calf, weaning to 8 month, 450 lb	0.50
Goats	
Doe and kid pair	0.24
Goat, mature, nonlactating	0.17
Kid, weaned	0.14
Horses	
Mature, draft	1.50
Mature, saddle	1.25
Sheep	
Ewe and lamb pair	0.30
Sheep, mature, nonlactating	0.20
Lamb, weaned	0.14

Source: Valentine, *Grazing Management.* (San Diego, CA: Academic Press, Inc.) 276–293.

Once you know the forage consumption of an animal and have estimated your forage production, you can calculate how many animals your pasture or range can support and for how long. Rough estimates of forage production are available from your UC Cooperative Extension Farm Advisor, or you can ask neighboring livestock producers (but make sure their herds look healthy). Ultimately, your own experience and monitoring of actual pasture or range use will give you the best information.

Climate, topography, and other factors complicate an area's grazing capacity. Table 11.2 shows the effects of slope and canopy cover (shading from shrubs and trees) on estimated grazing capacity for three climatic areas in California.

Most pastures' production can be greatly improved if you follow good land and animal management practices. Fertilization, drainage, rotational grazing, regular clipping, weed control, pasture renovation, and reseeding all improve pasture productivity. For example, a mature horse requires 2½ to 3½ acres of dryland pasture, versus 1 acre of irrigated pasture.

To determine the number of AUs per acre per year that your range will carry, divide the value for your climatic zone, slope, and canopy percentage by 12. For example, table 11.2 shows that a northern California range with a canopy cover of 0 to 25 percent and a slope of under 10 percent can provide 3.5 AUM per acre. Dividing by 12 annualizes the monthly figure, resulting in about 0.3 AU per acre per year.

3.5 AUM/acre ÷ 12 months/year = 0.3 AU/acre

How many acres are needed to support 1 AU?

1 acre ÷ 0.3 AU/acre = about 4
(rounded up from 3.3) acres for 1 AU

Once you have chosen the type of animal you wish to raise and have figured out how many animals your pasture can support, you can use several management tools to improve either the production of the pasture or the grazing efficiency of your livestock. Pasture production (yield or quality) can be improved with fertilization, weed and brush control, irrigation and renovation and reseeding. Before you undertake range or pasture improvement, prepare a cost and benefit analysis.

Pastures cannot be grazed continuously, without a period of rest when the plants can regenerate themselves, or the result will be overgrazing. Overgrazing kills plants and can lead to bare soil that is susceptible to erosion. Overgrazing is not a function of the number of animals on a pasture, but of how much time those animals are allowed to graze. Plants can be severely defoliated once and then recover if they are given time to regenerate and their nutrient requirements are adequately met. A plant's ability to store nutrients in its roots or to grow requires adequate leaf exposure to sunlight. Grazing animals are highly selective in their choice of which plants in a pasture they will eat, even though the plants they avoid may be equally nutritious. This can cause overgrazing of a few plants in a pasture.

One of the deleterious effects of overgrazing is the degradation of water quality due to sediment erosion from bare ground. Also, improper use of land alongside streams may impair water quality. There are regulations that govern aspects of the use of irrigated lands or pastures, many of them implemented by local watershed groups.

You can control overgrazing by means of a variety of management schemes that will result in increased grazing efficiency: more plants grazed to proper height without overgrazing of a select few.

Time-controlled grazing (also known as intensive grazing management, high-density/short-duration, or mob stocking) divides the pasture area into multiple paddocks. The grazing animals are heavily stocked in a single pasture and then moved to a new paddock. Timing for

Table 11.2. Estimating grazing capacity of dryland range in three California climatic zones with various slopes and canopy cover

		AUM per acre by slope class			
Area	Canopy cover (%)	Under 10%	10–25%	25–40%	Over 40%
Southern California (10" precipitation)	0–25	0.7	0.4	0.3	0.1
	25–50	0.4	0.3	0.2	0.1
	50–75	0.2	0.1	0.0	0.0
	75–100	0.1	0.0	0.0	0.0
Central Coast, Central Valley foothills (10"–40" precipitation)	0–25	2.0	0.8	0.5	0.3
	25–50	1.5	0.6	0.4	0.2
	50–75	1.0	0.4	0.3	0.1
	75–100	0.5	0.2	0.2	0.1
Northern California (over 40" precipitation)	0–25	3.5	1.3	0.8	0.5
	25–50	2.8	1.0	0.6	0.3
	50–75	1.8	0.7	0.5	0.2
	75–100	0.9	0.3	0.2	0.1

Source: McDougald et al., Davis, CA: UC Davis Agronomy and Range Science, 1991.
(The figures in this table are not for rotational grazing.)

when to move the animals is based on the plants' growth cycle. This method allows more stock to be carried on a pasture without overgrazing it, in contrast to either continuous set stocking or other rest/rotational methods. The total carrying capacity as calculated from the soil type, canopy cover, and topography for the pasture remains unchanged, though, regardless of the management scheme. Under conventional set stocking or non–time-controlled rest/rotational grazing schemes, managers stock less than the pasture's total carrying capacity to prevent overgrazing.

Time-controlled grazing is complicated. For advice on how to use it, consult your local UC Cooperative Extension Livestock Farm Advisor. If incorrectly managed, high-density–short-duration grazing can destroy a pasture very quickly. If you run multiple species of livestock, the species or animals with the highest-quality diet requirements should be rotated through a pasture first.

Housing and space requirements. Just as humans require clean housing protected from the elements, so do animals. Like humans, adequate room to exercise is of paramount importance to the health and vigor of the animal. Size, age, sex, species, and number of animals will influence shelter and space requirements. A catalog listing USDA plans for farm buildings and other farm structures is available for reference in some county UC Cooperative Extension offices. Volunteers in 4-H may also have this information available for you. Individual plans may also be available online as free downloads from Iowa State University (http://www.public.iastate.edu/~mwps_dis/mwps_web/frame_p.htm), and more complete books are available for purchase as well (from http://www.mwps.org). Contact your local UC Cooperative Extension Farm Advisor for assistance once you decide on the type of livestock you wish to raise. Contact your city or county building department to determine whether the design meets codes and whether you need to obtain any permits before you start construction.

Table 11.3 lists guidelines for suggested minimum space requirements. Common sense should direct you to allocate enough space so your animals do not harm themselves. Too much space, on the other hand, can make catching an animal difficult and may limit your potential production or profit. In warm climates or during the summer, plan on using the upper limits of space. Extra space will help the animals dissipate heat and remain cooler. This is especially important if you raise hogs. Watch for future California regulations affecting space requirements prompted by passage of Proposition 2 Standards for Confining Farm Animals in 2008.

Table 11.3. Space requirement per individual for various livestock and poultry*

Animal	Floor space (sq ft)	Yard space (sq ft)
Cows		
2 years or older	40–50	300
Cattle, yearling, finishing	30–40	125–200
Calves, 350-500 lb	20–30	100–175
Cows, pregnant	100–120	1–2 acres
Herd bulls	100–150	1–2 acres
Goats and sheep		
Ewes, dry	16	16–20
Goats, dry	16	200
Ewes with lambs, does with kids	20	30
Stud rams or bucks	20–30	30–60
Poultry		
Chickens, laying*	*	NA
Chickens, broilers	1	NA
Ducks	3–4	4–5
Rabbits	3–4	NA
Swine		
Sows, before farrowing	15–20	15–20
Sows with pigs, gilts	48	48
Sows, mature with pigs	64	64
Herd boars	15–20	15–20
Swine, growing-finishing, under 75 lb	5–6	6–8
Hogs, growing-finishing, 75–125 lb	6–7	7–9
Hogs, growing-finishing, over 125 lb	8–10	8–10

Source: M.E. Ensminger. *The Stockman's Handbook.* 5th ed. (Danville, IL: Interstate Printers and Publishers, Inc. 1978).

* At the time of this writing, parties involved in the adoption of Proposition 2 in California are still seeking legal interpretation of the new law. Still to be determined are (1) which specific housing systems are legal, and (2) which space allotments are legal.

Health care. Though you can administer routine health care yourself, you need a veterinarian to provide preventive care and to treat diseases, parasites, injuries and illnesses. Health care and the availability of veterinarians varies greatly for different species of livestock. Routine health care is more complicated and expensive for large animals. Your local veterinarian can give you a good idea of routine health care your planned livestock ranch will need and an estimate of costs.

Managing animals. Your own knowledge of your physical limitations will help you narrow your choice of what livestock to buy and make your project both successful and enjoyable. The larger the animal, the more physical strength is required of the handler. Managing a 600-pound steer is much more difficult than managing a 10-pound rabbit. Male animals tend to be more aggressive and larger than females or neutered males. You must be available to feed, protect, observe, and treat your animals several times every day. You will spend time on record keeping, learning new information, and attending field days and meetings.

Marketing. Study auctions, cooperatives, and other buyers; learn what time of year to sell; decide whether you will harvest before you sell. What are the laws with regard to harvest? Will your animals require USDA inspection? Where will you sell? Will it be through niche or farmers markets? Do you have transportation available? How much will transportation cost? For more information, see *Selling Meat and Meat Products* (ANR Publication 8146, http://anrcatalog.ucdavis.edu/pdf/8146.pdf) and the "Resources and References" section at the end of this chapter.

Animal identification. Although animal identification is not required, it facilitates good management and marketing. With good records on how an animal or its offspring perform, you can monitor animal performance and select and breed better animals. Also, some marketing programs require animal identification so that producers can keep records and verify whatever the particular marketing program requires. Food safety is enhanced further when an animal can be traced through the production system, assuming that good quality assurance practices specific to each species are followed. Everyone raising sheep and goats in California is required to have official identification issued by the USDA in order to sell them at a public auction. You may obtain official tags from the USDA/APHIS/VS office at (916) 854-3900. If you apply for an official ID, you must retain records on those sheep or goats for 5 years, including the date you identified the animals and the identification numbers applied, such that animals may be traced back to their flock or herd of origin.

POULTRY

Most poultry and egg production in California is on large, specialized ranches. Feeding, watering, and egg collection are usually mechanized. Just a few companies process most of the fryers found on the commercial market. Turkey production is concentrated on a few large ranches in the San Joaquin Valley.

A number of small-scale producers have been able to survive and compete in the specialty bird market. Most sell directly to consumers. Carefully investigate the market for specialty poultry before you start production. A small distributor or processor may be looking for a certain volume of a specialty item; if more is produced, the price will quickly fall and so will your profit.

A few types of specialty poultry and exotic birds that small farmers successfully raise and market are chickens, ducks, quail, squab, turkeys, partridges, pheasant, and guinea fowl.

Chickens. Many small ranchers can earn additional income by selling eggs directly to consumers or by producing eggs that command higher prices, such as brown-shelled eggs, fertile eggs, or free-range eggs. None of these products is considered by scientists to be nutritionally superior, but some consumers are willing to pay a premium for them. Freshness, while not assured in these types of eggs, may be a factor in consumer decisions.

Recently there has been an increasing market for dark-feathered (usually red or black) meat chickens and White Silkie chickens. They are usually marketed live to small poultry markets or directly to consumers. Dark-feathered meat-type stocks are available as day-old chicks. White Silkie chicks usually are produced on the ranch where they are grown. These chickens command premium prices and can be profitably produced on a small ranch.

Ducks and geese. Most ducks grown in the United States are the Pekin breed, which is used extensively for meat. The market for geese is limited, but they are popular on farms for eating weeds, as watch animals, and for home-produced meat and eggs. The duck-raising business is dominated by a few very large producer-processors. There are several alternatives that are much more likely to be profitable for small ranches.

The production of duck eggs (usually fertile) has proven to be a profitable enterprise on several California ranches. Any breed can be used, but the most prolific layers are Khaki Campbells and Indian Runners. The eggs are often sold fresh, salted, or pre-incubated for about 18 days. All of these are popular with California's large Southeast Asian population. The adult birds are often sold live at farmers markets or to small processors, and bring a reasonable price for meat. Muscovy ducks are raised for meat. They are often crossed with other domestic ducks (often Pekin). The crosses are sterile and will not reproduce. The Muscovy has a much leaner carcass than other breeds of duck, and the meat sells for a higher price.

Japanese quail. Japanese quail have been raised commercially in the United States since about 1960. They were first imported for use as game birds, but that venture was not very successful. Now they are sold for meat and egg production. The eggs are often marketed at farmers markets and through specialty food stores. Meat is sold through Asian markets and to high-end restaurants. The key to success is careful planning and development of a reliable market.

Quail production usually requires the production and incubation of fertile eggs, since day-old chicks are not widely available. The birds mature at 5 to 6 weeks of age and are prolific layers. About 12 ounces of feed is required to produce a dozen eggs.

The price for meat birds depends on quality, size, and demand. A jumbo strain, which at 3.5 to 6 ounces grows to about twice the size of the original imports, has been developed. Quail are raised in paired cages, group cages, wire-floor pens, or litter-floor pens. Wire-floor units are often used because separation of the birds from their manure reduces disease problems.

Squab. A squab is a young pigeon. Squabs can be a source of income with good management and a good market. Investigate the market before you seriously consider production. Simple housing consists of pens with feeders, waterers, roosts, and nests. These pens are often open on three sides and screened with poultry netting. The nests are built on the solid back wall. Normally 4 to 5 square feet is allowed per pair, with up to 20 pairs per pen.

A pair of squabbing pigeons incubates and feeds young squabs. With good management a squabbing pair will produce 10 to 13 squabs per year. Squabs are processed (killed and dressed) at 28 to 30 days of age, when the pinfeathers under the wing are in their sheaths and not quite open.

A unit of 500 squabbing pairs requires about 30 hours of labor per week, and, like all poultry raising, this is a 365-day per year job. Squab raising is compatible with other ranching ventures. It provides regular income, and the unique taste and texture of the meat makes it very marketable with potentially good returns

Game birds. Some farmers raise game birds, principally pheasants, partridges and quail, to sell to shooting clubs, wildlife managers, and landowners to increase resident populations. Other game birds, including wild turkeys and guinea fowl, are sold mainly for meat.

There is an expanding market for meat from game birds, and most of the work can be done during the summer. But the public often has a negative attitude toward hunting and pen-raised game birds, and their management is very time consuming.

BEEF CATTLE

Enterprise selection. Beef cattle can be raised on a small farm. There are many potential beef cattle enterprises for you to consider

- **Cow-calf.** You can sell the calf at weaning (at approximately 400 to 500 lb), retain the female calf as a breeding animal (heifer), or sell it later as it feeds and puts on more weight eating grass. Breeds that do best are specific to an area, with *Bos indicus* (Brahman or Zebu) and their crosses more appropriate for very hot regions.

- **Stocker.** You buy 400- to 500-pound weaned animals in early winter, run them on dry range, and sell them in May or early June. If you have irrigated pasture, you can buy weaned animals in late winter or early spring and sell them in late summer or early fall. Typical sale weights are 700 to 800 pounds. You should watch the market to determine the best time to sell. There are times it can be more profitable to sell at a lighter weight.

- **Direct marketing of beef.** You finish a raised or purchased animal to a harvest weight and then sell the meat to consumers or use it for your own consumption. The finishing period is used to get animals up to around 1,000 to 1,100 pounds live weight. The finishing feed can either be all forage or a mix of grain and forage. Meat sold by the quarter or half (side) must be harvested at a USDA-inspected site. Any questions regarding the selling of meat can be directed to the California Department of Agriculture's Meat and Poultry Inspection Service, telephone (916)

654-0504. A list of USDA-inspected facilities can be found online at http://ceplacernevada.ucdavis.edu/Custom%5FProgram550/USDA_Inspected_Harvesting_Sites.htm.

Infrastructure. There are several elements of infrastructure that need to be installed before any animals arrive on your property:

- **Fencing.** Do you have a perimeter fence and is it in good shape? Many places have barbed wire, smooth electrified wire, or field fencing. You should check the entire length of your fence to make sure your animal can be contained on your property. Interior fences can either be permanent or temporary, using portable electric fencing materials.

- **Livestock water.** Are there water points for cattle? A cow will drink 15 to 20 gallons per day during the hot part of the summer. Troughs can be either metal or plastic. It is advisable for you to develop several water points on your property to improve livestock grazing distribution throughout your entire property. If you live where it freezes in the winter, make sure you have a way to provide regular drinking water in all seasons.

- **Corrals and working facilities.** These are needed for general health management (e.g., worming and vaccines), treatment of sick animals, and loading or receiving animals. For example, a squeeze chute is ideal for treating animals. There is nothing more frustrating than to have an animal with a weepy eye (pinkeye) that needs treatment, but to find yourself unable to take care of it because you have no way to restrain the animal. You may also be able to get by with a head gate.

Health and nutrition considerations. Beef cattle require annual vaccinations for clostridial and respiratory diseases, and you should deworm cattle at least once a year. Consult with a local veterinarian to set up an annual vaccination and deworming schedule. The University of California has some sample health schedules for different classes of cattle and horses for the northern Sierra foothills available online (http://groups.ucanr.org/sierrafoothill/SFREC_Animal_Health_Programs/). Both annual and irrigated forages grow most rapidly in the spring. During this time of the year, the grass can start to mature and form seedheads. Grazing cattle can sometimes get pinkeye infections from irritation to the eyes from seedheads. Check with your veterinarian to determine the best course of treatment.

Beef cattle should have access to feed and water at all times. Annual dryland range will be high quality (exceeds protein and energy requirements) from mid-March through mid-May. With seedhead formation, protein levels begin to drop. From the hot part of the summer into fall, there may not be enough protein in the forage to satisfy a cow's nutritional requirement, so you should supplement for protein during this time of the year. Contact your local livestock UC Cooperative Extension Livestock Farm Advisor for protein supplementation options.

Irrigated pasture will more than meet the nutritional requirements of the cow during the irrigation season. To prevent soil compaction, do not graze animals on pasture while it is being irrigated. After irrigating, they should be kept off the pasture for at least 24 hours.

Many areas of northern California are selenium deficient. A lack of selenium can cause general unthriftiness of the animal, reproductive problems, and white muscle disease. You can check the Trace Minerals for California Beef Cattle Web site (http://animalscience.ucdavis.edu/MineralProject/) to see whether your area is deficient in selenium.

A trace mineral supplement should be available at all times. This can be in the form of a block or a loose mineral. Make sure you move the mineral supplement along with the cattle if you move them to a new paddock or pasture.

In some months of the year, usually December through February, pasture or forage feed may be lacking. During these times you will need to provide supplemental feed in the form of hay, haylage, or silage to keep your cows in good nutritional condition. It is important to have your cows in a good nutritional state at calving. This will shorten the recovery period of the reproductive tract after calving and make it easier to get the cows rebred.

Reproduction. Farmers have two options for stock reproduction: use of a bull or artificial insemination. For a farmer with a small acreage, it makes the most sense to use artificial insemination. Artificial insemination requires special skills, but you can acquire those at a course sponsored by an agricultural college or university or you can ask other local cattle producers to help.

DAIRY COWS

In general, dairy cattle are not suited to commercial small farm enterprises for the purpose of selling milk or processing dairy products in California. The reason is that both milking facilities and milk processing are regulated by the state, and the cost of developing approved facilities, together with inspection costs and licensing fees, make it difficult to make a profit on a very small scale. In addition, to keep production costs low, the small farm needs to have adequate land on which to graze the cattle and from which to harvest additional forage to feed in the winter.

In California there is no regulatory exemption for small numbers of cattle or goats. If you only own one cow or just a few goats and offer any milk for sale to the public, it must either be produced in a licensed milking facility and processed on your farm in a licensed processing plant or sold to a licensed processor and transported to that plant via a licensed milk handler. *A small farm cannot legally sell milk to neighbors or at farmers markets without this licensing.* In addition, commercial milk processing cannot take place in a kitchen that is approved for other value-added or food processing, such as for making jams and jellies or barbecue sauce. Milk processing has to be the only activity that occurs within the facility.

There are small-scale cheese makers in California and other states that operate small-scale dairy businesses with fewer than 100 cows. Most of these farms have a sufficient land base to provide grazing and produce this value-added product. Artisan cheese can bring higher returns to an operation than can commodity sales. However, these farms do not typically rely on the dairy for 100 percent of their income.

Manure management has become a critical aspect of operating a dairy farm. If you decide to establish a commercial dairy enterprise, you will need to have a permit from California's State Water Quality Control Board. The permit entails following regulations, keeping records, and submitting documentation to a Regional Water Quality Control Board. Although small commercial enterprises may have less paperwork to complete than large dairies, it is still a required activity; failure to comply will result in fines and other penalties.

If the small farm operator's intention is to produce only enough dairy products for their own family use, they do not have to meet any regulatory requirements except for animal health vaccinations and ownership identification requirements as required by the state veterinarian.

Additional information regarding animal care and space requirements is available online (http://www.vetmed.ucdavis.edu/vetext/inf-da/INF-DA_CAREPRAX.HTML).

SWINE

Farmers interested in raising a small number of pigs on their farm can find appropriate information in *Handbook for the Small-Scale Pork Producer* (ANR Publication 21435). Subjects covered include site selection, housing requirements, health, marketing, and breed associations. You can also access the information online (http://www.vetmed.ucdavis.edu/vetext/INF-SW_CarePrax.html).

GOATS

Goats are increasingly popular on small farms. They are adaptable to many climates, require only simple housing, and are small enough to handle and transport easily. They are the most populous animal species on farms around the world because they provide food, fiber, and hides for millions of people and they control unwanted vegetation. Start with quality breeding stock to build your herd. The following information will help you understand the needs associated with raising goats.

Shelter. Goats do not like rain and need cover. Dry, warm shelter is especially important for young animals. Hay and grain for feed require a large storage area. A grain tank and a barn are necessary if you want to get the best prices by purchasing feed in bulk.

Feedlot or pasture? Farmers raise goats successfully in both feedlot and pasture situations. Goats are natural browsers, not grazers. Forage grazing helps reduce your feed bills and provides the goats with exercise and fresh air. For milk production and growing meat animals, high-quality hay and grains or other supplements are needed for much of the year. Understanding the animals' nutritional needs for best management practices is both a science and an art.

Fences. Goats jump and crawl through fences and can therefore be hard to contain. Fences for even very young kids should be high enough to prevent them from jumping out by accident. A typical height for fences to contain large kids and dry does is 5 to 8 feet. Electric fences can be lower (as low as 4 feet), but you will need to train the goats to respect them. Electric fences also provide for some protection from predators.

Goats need to be protected from wild and domesticated predators. As mentioned above, electric fencing is useful. Many producers also use guard animals to help safeguard their goats.

Knowledge and market access. Most small farms use a combination of marketing strategies to get the best prices for their products, using direct marketing and wholesale channels. Cheese can be sold at farmers markets; milk and meat are more difficult to sell directly (see below for regulations). Distance to market is becoming more of a cost factor as fuel prices increase. Collaboration with others in your area may help you offset higher costs.

Health issues. Prevention is the best medicine. As in any livestock operation, when you raise goats for commercial production, health issues can be exacerbated by poor management practices. However, even the best managers may encounter infectious diseases, poor-quality feed, toxic weeds, breeding problems, and hoof rot. Knowledge of potential problems can help you prepare to deal with whatever may come your way. There is a great deal of information available about raising goats and many anecdotal remedies. Make sure to rely on reputable sources of information and work with a veterinarian and a UC Cooperative Extension Livestock Farm Advisor. It is also a good idea to find an experienced operator to be your mentor.

Culling weak, poor-quality, and low-producing animals is another management tool for health issues and is essential for production economics. Beginning operators' experience indicates that they tend to retain animals that do not improve their profitability, even though they understand the need to cull more heavily. For example, producers may feed an animal that they know will not produce a return, even though feed costs are among their highest expenses.

Goats for meat production. In the United States, Spanish, Boer, Kiko, and Myotonic goats (Tennessee wooden-leg or Fainting goats) are the most common breeds used for meat. Some dual-purpose (meat and dairy) goats are also used and include the Nubian, Kinder, and Angora. The Pygmy, West Africa's favorite meat goat, is sometimes used for meat. Standards have been developed for the Boer and Kiko breeds, and they are the most sought after.

Small farms have begun to raise meat goats in response to a burgeoning demand from specialty products and ethnic markets. Raising goats for this identified niche market raises one's hopes for profitability, but it is not a sure thing. Competition is fast becoming a reality, but opportunities do still exist. It is best to start small and develop your market. You can explore some value-added attributes, such as direct marketing and special processing to qualify for religious certifications.

Dairy goats. For many years there has been a steady market for goat milk due to its digestibility among the lactose-intolerant population. The increasing popularity of specialty cheeses has heightened industry awareness of goat milk and its usefulness in making unique soft (chevre) and aged cheeses. Many types of cheese are made from goat milk, and its popularity continues to grow.

Small dairies are a challenge to operate because of the 7-days-a-week milking schedules and other demands. Commitment by the producer is paramount, especially because outside labor can be expensive and therefore hard to justify. Creative approaches are necessary to meet the labor demands: for example, providing housing, sharecropping, and using family labor. There are value-added opportunities: many small farms add a cheese room and make homestead cheeses and yogurts to obtain the premium consumer dollar. However, production and marketing of these products will add to your work and labor needs. Selling wholesale milk to a processor requires that you keep many more animals than a homestead cheese business does, but it also eliminates the need to produce and market value-added products.

These guidelines are very general, but you may find them useful: if you have only occasional hired labor, 150 to 200 milking does is a manageable herd; when you add a full-time employee, the herd must be larger (800 to 900 does) in order to be profitable. At that point you are dependent on hired labor because the producer alone cannot milk that number of does each day. You can find a dairy goat cost of production study online at http://coststudies.ucdavis.edu.

Setting up a new goat dairy is not a one-step process. Seldom can you find a whole herd for sale, so you will need to develop your own herd, and that entails buying animals of various ages and breeding and raising young does. Building the initial herd can take 2 to 4 years. These first years are a financially challenging time: you will need to buy feed and equipment, but you are not likely to sell any milk during the first 1 to 2 years. When you first start milking, you may not have enough milk to have it profitably processed. It is typical for a farmer to use the first year's milk for raising kids to build the herd. Learn the conformation of animals before you buy dairy stock, especially the mammary system and leg structure. Caprine arthritis encephalitis (CAE) is a viral infection of goats that may lead to chronic disease of the joints. When building a herd, it is best to avoid goats that carry this virus and to buy only tested does. Buy high-quality bucks: your future herd will have their genetics. Of the five common dairy goat breeds, Alpines are best known for their hardiness. A 2005 survey identified Alpine as the breed of choice among commercial producers.

You will need to work with county and state health department officials because different agencies regulate milk and cheese processing facilities, and the retail trade. All milking facilities are required to be inspected. Homestead cheese processing rooms are also registered and inspected monthly. Sanitation is paramount for food safety, best management practices, and the success of your business. Problems with food safety and food recalls can be devastating for a small farm business. Risks must therefore be identified, planned for, and managed.

SHEEP

Sheep are very well suited for the small farm, especially if the farm has adequate pasture or rangeland to meet most of their nutritional needs. Sheep are often used to graze crop residues, but you must take proper care, since some crop residues are high in compounds, like nitrate, that can be toxic. Feed expense, as with any livestock enterprise, is typically the highest cost associated with a farm flock. Being smaller animals than cattle, sheep require less land, physical strength (for handling), and capital expenditures for buildings, fences, and working facilities. Sheep do require more protection, however, from domestic dogs and wild predators. Good information on guard dogs is available online (http://www.aphis.usda.gov/wildlife_damage/dogs/information.shtml). Electric fences and guard animals (most often guard dogs) help protect farm flocks. Labor becomes intense at least two times a year: during lambing and (for wool breeds) at shearing. By learning to shear sheep, you can both reduce the cost associated with your own flock and add substantial seasonal income from shearing others' sheep, with a minimal investment in equipment (usually less than $2,000). A few breeds of sheep actually produce no wool (hair sheep) or shed their wool annually, which may be an important consideration, given the increasing difficulty of finding qualified sheep shearers in the United States, including in California.

The small farmer has several decisions to make about adding a sheep enterprise to his or her operation. Most sheep are raised for both their meat and their fiber. Different breeds have evolved with an emphasis on wool production (e.g., fine, long, medium, and coarse or carpet wool and naturally colored wool) and on meat production. Meat breeds are often referred to as terminal sire breeds, as the males (rams) are often mated with a female, or ewe, from one of the wool breeds to improve the resulting lambs' carcass quality. Some breeds are known as dual purpose, meaning that they produce a good wool clip as well as a good meat carcass. Dairy sheep breeds are more common in Europe, but their numbers are growing everywhere in the United States. Sheep's milk is about 50 percent higher in total solids than cow's or goat's milk, and high-quality cheese made from sheep's milk commands a premium price. Sheep produce less milk than goats do, and production costs are higher. About 49 different breeds of sheep are commonly found in the United States. Some are more readily available than others for use in a small farm flock, including Suffolk, Dorset, Hampshire, and many crosses.

Hand in hand with the choice of breed for a sheep enterprise is the small farmer's choice of the actual type of operation. In some areas, lambs can be weaned from the ewe and finished for slaughter straight from the pasture or rangeland. In other areas, the lamb is weaned and then sold as a feeder lamb. The former practice lets you address niche markets such as grass-fed or organic markets, provided that you follow strict rules and meet certification criteria. With the latter practice, growers usually sell the feeders through an order buyer or an auction, or direct to a feeder-packer. Sometimes, when growers have available feeding area or pasture and a cheap source of feed, they may feed the feeder lambs themselves and eventually sell finished lambs. Direct marketing of meat requires that the lamb be processed at a federally or state-inspected facility. Some farm-flock producers specialize in wool for hand-spinners and weavers, and many process the wool into yarn or other value-added products. Some of these producers further specialize in raising natural-colored fleeces or raising minor breeds like the Navajo Churro, whose wool is used for the famous Navajo blankets.

Purebred or registered sheep production is another highly specialized operation that produces seed stock for show and sale. Several purebred producers of terminal sire breeds sell rams to commercial flocks. Some producers specialize in marketing project or club lambs for 4-H and Future Farmers of America (FFA) youth projects. Like goats that are used for brush control, sheep flocks are now being hired out to graze for weed control or floor management of dormant vineyards or orchards. They are also used similarly in some agro-forestry operations to decrease weed competition in reforestation efforts. Reduction of fine fuel for wildfire prevention is another type of grazing service that has been made possible by the portable electric fence. Often, sheep used in these grazing services are promoted as environmentally friendly, since they recycle the weedy species and reduce the land manager's dependence on fossil fuels and pesticides.

Before deciding whether to raise sheep as a small farm effort, we encourage you to attend workshops offered by UC Cooperative Extension and to learn more by studying the references and resources on sheep provided at the end of this chapter. Consult with the local UC Cooperative Extension personnel and with local large livestock veterinarians before you embark on a sheep enterprise. If you are new to sheep raising, start small and gradually expand your operation as your knowledge and experience grow. Raising sheep can be an enjoyable and rewarding enterprise.

BEEKEEPING

There are various ways to make money from keeping honey bees. Most beekeepers rely on honey and beeswax production and commercial crop pollination. A few raise queens and bulk bees for sale. Costs to start a beekeeping business are not particularly high compared to many small businesses, and a well-planned and managed operation can be profitable (see sidebar).

Beekeepers own, rent, or find rent-free apiary locations where their bees can forage for food without becoming a nuisance to humans or livestock. Beekeepers must manage their colonies to the benefit of the bees and in compliance with existing state, county, and municipal beekeeping ordinances. In some areas, hives can be located on one spot permanently. In many areas, though, a permanent location would not adequately meet the needs of the bees so the beekeepers relocate their apiaries a number of times each year.

Bees are essential for pollination of more than 50 California crops, so farmers who do not keep their own bees rent them from a beekeeper. Colony rental prices vary according to supply and demand. Summer vine seed pollination in California is extremely inexpensive for growers because the beekeepers are searching for safe havens where they can put their bees and where the bees might get a little food. Rental rates are as much as 10 times higher for almond orchards, which need to be pollinated at a time of year when bee supplies barely meet demand.

Part-time beekeepers operate 50 to 450 colonies, while full-time beekeepers operate 1,000 to 4,000 colonies. One knowledgeable beekeeper is needed to handle every 500 to 1,000 colonies.

Beekeepers need a place to store and repair equipment, mix the bee feed and antibiotic treatments, and extract and handle honey. Some people use a garage or toolshed, but a larger facility usually is required. Many beekeepers rent a building. Others build a honey house. Before building a facility, it's a good idea to visit other beekeepers and make note of features and techniques they use to make the handling of equipment and honey efficient and sanitary.

Beekeeping can sound deceptively simple, but in fact beekeeping is a form of animal husbandry that involves providing feed when nectar and pollens are lacking, preventing infections from various microbes, dealing with two well-established parasitic mites, and reducing the influence of Africanized honey bees. Before you try to keep bees commercially on your own, you should gain experience working with a commercial beekeeper for one or more seasons.

LLAMAS AND ALPACAS

Llamas and alpacas are making a growing impact on the small farm. Their rise in popularity is due to their mild personality and general ease of management. Excluding animal purchase costs, maintenance expenses for llamas and alpacas are generally low compared to those for other livestock, and the profit potential is attainable with a relatively small number of animals. It is important to note that although these animals are often purchased or raised as companion animals, they are considered livestock, and you need to check local zoning regulations before purchasing them.

Llamas and alpacas are among the oldest of domesticated animals. Microsatellite DNA research confirms that they are domesticated from two species of wild camelids (llamas from guanacos and alpacas from vicunas) that are still present in the South American Andes (Kadwell et al. 2001). Most of the world's population of all four species of camelids is still in South America. Because of the small number of these animals in the United States, registered purebred lines are the most common enterprise. This means that your initial animal purchase will be fairly expensive, especially for alpacas. With such a large capital investment, any purchase should be from a reputable herd that follows proper nutrition and herd health protocols for the area.

Llamas are much larger than alpacas, making them easy to distinguish from one another. Mature llamas weigh between 250 and 450 pounds, while alpacas range from 100 to 175 pounds. Another distinguishing feature is that the llama's back tends to be straight and the alpaca's back curves slightly upward. The two animals have many other characteristics in common, such as

ESTIMATED INVESTMENT NEEDED FOR A 1,000-COLONY BEE OPERATION

Item	$
Hive equipment	
1,000 bottom boards @ $8 each	8,000
1,000 covers @ $8 each	8,000
2,000 deep boxes @ $12 each	24,000
20,000 deep frames @ $0.50 each	10,000
20,000 deep foundation @ $0.90	18,000
1,000 medium-depth boxes @ $8 each	8,000
10,000 medium-depth frames @ $0.50 each	5,000
200,000 frame eyelets @ $17 per 5,000	680
6,000 metal rabbets @ $0.13 each	780
1,000 queen excluders (optional) @ $5 each	5,000
20 fume boards @ $10 each	200
1 bee blower (optional) @ $450 each	450
75 gallons paint @ $25 per gallon	1,875
1 staple gun and compressor @ $750	750
1,000 packages of bees @ $40 each	40,000
Honey-handling equipment	
Automatic uncapper	2,500
Frame conveyor	700
Conveyor drip pan	275
Cappings melter	2,000
Extractor	6,000
Settling tanks (each)	275
Spin float (replaced melter)	3,000
Honey sump	800
Honey pump	2,500
Flash heater (optional)	1,000
Barrels (each)	
new	20
used	10
Barrel truck	450
Hand truck	370
Glass or plastic jars (if not selling bulk)	18,000
Bottling device (if not selling bulk)	1,000
Vehicles	
Flatbed truck (each)	20,000
Bee boom (each, mounted)	4,000
Forklifts (each)	
new	20,000
used	12,000
Pickups (each)	
new	35,000
used	20,000
Warehouse	10,000
Land (5 acres @ $4,000 per acre)	20,000
Rent (house and shop, per year)	20,000
Labor	
Self (range)	50,000-60,000
Help, full time, each (range)	20,000-30,000
Help, part time, each	2,000
Overhead	
Utilities (per year)	2,400
Insurance	(varies)
Workers' comp., health insurance	13,000

their split-toed ungulate feet. They also share the distinct camelid trait of being able to spit as either an offensive or defensive gesture.

In the United States, both llamas and alpacas provide sources of income through sales of the animals and through fiber production. Sales of purebred lineage make it possible to create markets that can provide a positive return on investment. Livestock shows have been used as a means of marketing top breeding animals. Breeder and registry information for both species can be found in the resources and references at the end of this chapter. A spring shearing can help to keep the animals cooler in the summer and provide an opportunity for economic benefit through the sale of fiber. Stiff hairs (also called guard hairs) found in the llama coat do not exist in the alpaca coat. This makes the alpaca's fiber finer than the llama's, and thus worth more. Unlike sheep wool, alpaca and llama fiber has no lanolin, so it does not need to be processed before spinning. Llamas are also successfully used to guard sheep from predators and as pack and draft animals.

Fencing to contain llamas and alpacas is generally not difficult to construct or maintain. In fact, good-quality fencing, at least 4 to 5 feet high, is most necessary to keep predators out, especially for alpacas. Exterior fences should be tight and permanent, much the same as for sheep. It is quite common for interior fencing to be electric for an intensive grazing system. A restraining pen is a necessity for vaccination and hoof trimming. Since many of these animals come from high in the Andes, they are not accustomed to the extremely hot and sometimes humid conditions common to many areas of North America. To accommodate for this difference in climate, both types of animal require a three-sided shelter, with ventilation, for protection from harsh winter and summer weather.

Producers often make nutrition for llamas and alpacas more complicated than it need be. These species developed in a relatively harsher feed environment than they experience in the United States. Compared to other types of livestock, they have a low protein requirement but also a relatively poor efficiency of production. Pastures used for feeding llamas and alpacas are most often adequate, for nutritional purposes, with the exception of mineral and salt concentrations that may be deficient in a particular area. In fact, pastures with a high clover content can exceed their nutrient requirements and cause health concerns. Dry matter intake should be figured at 1.5 to 2 percent of the animal's body weight. Poisoning from weeds and some forages, such as oleander and, especially, plants with the ability to accumulate nitrates, is a potential problem that you should watch for and deal with as needed. Feed requirements for mature and idle adults are fairly low. Endophyte-free hays made up of grass species (orchardgrass, ryegrass, etc.) are more suitable than legume hays, such as alfalfa, when pasture is unavailable. Feed requirements do increase for lactating and growing animals, and you need to account for that. Contact your local UC Cooperative Extension Farm Advisor or veterinarian to find out about pasture mixes that are suitable in your area, possible local mineral deficiencies, and toxic weeds that may be present in your area. You should also work with a local veterinarian to set up an annual vaccination and deworming protocol.

WASTE MANAGEMENT

One of the by-products of animal agriculture is animal excreta. Feces and urine (manure) contain organic matter, nutrients, and pathogens. Nutrients include macro elements (those present in larger amounts): nitrogen, potassium, phosphorus, calcium, sodium, chloride, magnesium, and sulfur. Other components of manure may include animal hair, wool or feathers, spoiled or wasted feed, and bedding. Bedding may contribute less, as much as, or more daily weight (on a dry basis) than manure, depending on the animal species and type of housing.

Source generation. The volume of manure generated daily is a useful value for determining the volume of material you will need to collect at regular intervals. Recent updates to Table D384.2 of the American Society of Agricultural and Biological Engineers provide current estimates for the volume of excreta generated from commercially grown livestock and poultry on a daily (dairy or beef cow, laying hen, sow) or per-cycle (feedlot calf, broiler, turkey, market or feeder pig) basis. You can use these values to calculate the storage volume capacity you will need to have available during a specified time period. Although the table presents data for nutrient content (nitrogen, potassium, and phosphorus), increasing improvements in diets fed to animals have actually reduced the amount of nutrient excretion. It is important to avoid any overfeeding of nutrients (protein, energy, minerals, salts) so as to minimize the impact of manure on the environment.

Facility wash water or runoff water may be another source of waste. Wash water may have been used to wash animals prior to harvest of food products or it may have simply been used to wash the animals. Water that has come in contact with animals contains some amount of manure, dirt, and hair, and the water must be managed in a way that minimizes potential negative environmental impacts. In addition, it is important to contain and manage rain runoff if manure constituents are present.

You need to manage facility water to minimize standing water, water runoff, and mud holes. Standing water and mud holes attract undesirable insects, are a potential odor source, and may impair animal productivity. Functional water faucets (that turn completely off) to minimize spillage from troughs, and functional (unblocked, properly draining) gutters to convey water away from animal enclosure areas are helpful in reducing unmanaged discharges to nonfarm waters.

Collection. Waste from livestock housing areas is collected at varying intervals depending on the species and method of housing. For some species (e.g., horses), collection is daily or more frequent, whereas for other species (e.g., poultry, llamas, alpacas, swine, steers, lambs) less-frequent manure collection is common. Manure moisture content, form, and nutrient concentration all influence how frequently it needs to be collected. Fecal matter that is higher in moisture content spreads out when deposited by the animal and will be odorous and attract flies. Rapid drying or removal of such manure is key to minimizing nuisance conditions.

Treatment. A number of methods are employed to treat manure. The main reasons for treating manure include nitrogen stabilization, nuisance reduction (i.e., odor and flies), volume reduction, and pathogen reduction. Composting is a commonly used treatment method for manure on small farms. It is effective and does not require heavy equipment to get started.

Good compost criteria include maintaining a moisture level of between 40 and 60 percent and a carbon to nitrogen ratio of 25:1 to 30:1, along with regular turning of the compost when temperatures drop. As long as windrow piles sustain high enough temperatures for a long enough time (131°F for 15 days), you turn the compost a minimum of five times, and you cure the piles for an additional 45 days, pathogen reduction should be significant. Prevention of the reintroduction of pathogens to manure piles (either from the addition of fresh manure or the presence of farm pets or wildlife) is important to keeping the reduced pathogen load low.

Transportation. The primary method for transporting solid manure on small farms is via truck. Larger facilities may convey liquid waste from one location to another through a series of irrigation district pipes.

Utilization. Manure is collected in solid, semisolid, or liquid form. Solid manure contains some bedding and soil. As a solid, it is stackable and may be spread by means of a manure spreader or dump truck. Semisolid or semiliquid materials are not stackable but are too thick to be pumped easily through a normal sprinkler system. Liquid manure has a high water content (greater than 95%) and usually is discharged to land through a standard irrigation system.

One key concern for small farmers is the need to minimize the spread of pathogens from animal manures to crops that will not be processed prior to consumption. Recently adopted safety guidelines for the production and harvest of lettuce and leafy greens (http://safeleafygreens.com/) identify currently accepted guidelines for minimizing the spread of pathogens.

Environmental considerations. The protection of surface water and groundwater sources, air emissions, and human health impacts all are important. All confined animal facilities in California must comply with Title 27, Division 2, Subdivision 1 of the California Code of Regulations. These regulations, established in 1984, prescribe statewide minimum standards for discharges of waste at confined animal facilities. In California, the owners of a single animal, fed in confinement (absent pasture) for 45 days out of the year, must comply with the regulation. All facilities, regardless of facility age or location in the state must comply. The regulation was established to protect both ground and surface waters. A separate section of the California Code of Regulations identifies nuisance issues. Ultimate regulatory responsibility for nuisance and water quality protection lies with the nine regional water quality control boards of the state. More information on water quality control is available online (http://www.waterboards.ca.gov).

The California Department of Fish and Game also has regulatory authority (Section 5650) and prohibits discharges to surface waters that are deleterious to aquatic habitat. Additionally, the Federal Clean Water Act has regulatory authority over discharges to surface waters from point sources (discrete conveyances) as well as from non-point sources (diffuse conveyances) to surface waters. Point source discharges of nutrients from animal facilities to surface waters are prohibited.

Total maximum daily load (TMDL) limits have recently been adopted for many surface waters through a process that identifies impaired waterways (303 (d) list) and implements plans to reduce that impairment. Livestock and poultry operations existing in areas where surface water impairments have been identified must manage manure to prevent any off-site release of manure or feed nutrients. You can check online (http://www.waterboards.ca.gov/centralvalley/water_issues/tmdl/impaired_waters_list/index.shtml) to determine whether a waterway near to your operation is impaired.

The other major federal regulation with oversight on livestock and poultry operations is the Coastal Zone Act. Its reauthorization amendments mandated in the 1990s that states adopt management practices for waste containment structures as well as for nutrient application.

Air quality regulations for small farms have not yet been developed by the California Air Resources Board or local air quality management districts. It is important, though, that you manage animals and animal manures to minimize particulate emissions (i.e., fugitive dust) and minimize emissions of ammonia and volatile organic compounds into the air. Feeding just to the animals' needs (and not overfeeding them with nutrients) is an important step in minimizing the volatilization of ammonia. Care and management of animals may be particularly important to minimize the generation of dust.

Although significant fines may be levied when off-site discharges occur, and you do need to take care to minimize adverse effects on groundwater, it is nuisance issues that typically result in regulatory challenges. Good neighbor relations begin with keeping nuisance issues (flies, odor, animal vocalizations) to a minimum. Flies and odors can be controlled to some extent through regular removal of daily accumulations of manure, minimizing contact of water with manure (check waterers daily so you can quickly correct any malfunctions), and rapidly drying the manure.

RESOURCES AND REFERENCES

Animal Care Series and Selected Resources

Animal care series publications are accessible as free downloads: http://www.vetmed.ucdavis.edu/vetext/animalwelfare.

Beall, G., S. Larson, A. Nelson, and R. Phillips, eds. 1992. Sheep care practices. Animal care series. University of California Cooperative Extension, Davis. First edition. Pp. 27. http://www.vetmed.ucdavis.edu/vetext/INF-SH_CarePrax.pdf.

Farley, J. L., and W. J. van Reit, eds. 1991. Swine care practices. Animal care series.University of California Cooperative Extension, Davis. First edition. Pp. 25.

Harris, L. J., and H. L. Tan. 2004. Selling meat and meat products. University of California Division of Agriculture and Natural Resources, Oakland. Publication 8146. Pp. 6. http://anrcatalog.ucdavis.edu.

Jensen, W., and J. Oltjen, eds. 1992. Beef care practices. Animal care series. Pp. 49. University of California Cooperative Extension, Davis. First edition. Pp. 49.

Stull, C. L., ed. 1998. Broiler care practices. Animal care series.University of California Cooperative Extension, Davis. Second edition. Pp. 28.

———., ed. 1998. Egg-type layer flock care practices. Animal care series. University of California Cooperative Extension, Davis. Second edition. Pp. 25.

———., ed. 1998. Turkey care practices. Animal care series. University of California Cooperative Extension, Davis. Second edition. Pp. 29.

Stull, C. L., S. Berry, and E. DePeters, eds. 1998. Dairy care practices. Animal care series. University of California Cooperative Extension, Davis. Second edition.

Stull, C. L., and R. de Grassi, eds. 2000. Goat care practices. Animal care series. University of California Cooperative Extension, Davis. First edition.

Beef Cattle

Drake, D. 2006. Fundamentals of beef management. University of California Division of Agriculture and Natural Resources, Oakland. Publication 3495. Pp. 140. http://anrcatalog.ucdavis.edu/AnimalScience/3495.aspx.

George, M. R., G. A. Nader, and J. R. Dunbar 2001. Rangeland management series: Balancing beef cow nutrient requirements and season. University of California Division of Agriculture and Natural Resources, Oakland. Publication 8021. Pp. 9. http://anrcatalog.ucdavis.edu/pdf/8021.pdf.

Stull, C. L., S. Berry, and W. Jensen. 2007. Beef Care Practices. University of California Division of Agriculture and Natural Resources, Oakland. Publication 8257. Pp.38. http://anrcatalog.ucdavis.edu/pdf/8257.pdf.

Van Riet, W. J. 1980. Beef cattle production in California. University of California Division of Agriculture and Natural Resources, Oakland. Publication 21184. Pp. 24.

Dairy Animals

Stull, C. L., S. Berry, B. A. Reed, and M. A. Payne. 2004. Dairy welfare evaluation guide. University of California Cooperative Extension, Davis. Pp. 40. http://cdqa.org/dw_eval_guide.asp.

Swine

Farley, J. L. 1987. Handbook for the small-scale pork producer. University of California Division of Agriculture and Natural Resources, Oakland. Publication 21435. Pp. 52. http://www.vetmed.ucdavis.edu/vetext/INF-SW_CarePrax.html.

Miller, R. F. 1977. A practical guide to swine nutrition. University of California Division of Agriculture and Natural Resources, Oakland. Publication 2342. Pp. 8.

———. 1980. Swine production. University of California Division of Agriculture and Natural Resources, Oakland. Publication 21169. Pp. 12.

Goats

Berg, J. 2005. Raising dairy goat kids. University of California Division of Agriculture and Natural Resources, Oakland. Publication 8160. Pp. 20. http://anrcatalog.ucdavis.edu/pdf/8160.pdf.

Bowman, G. 1999. Raising meat goats for profit. Bowman Communications Press, Twin Fall, ID. Pp. 256.

Davy, J., et al. 2010. Sample costs for a goat meat operation in Northern California. University of California Cooperative Extension, Davis. Pp. 15. http://coststudies.ucdavis.edu/files/meatgoatncal2010.pdf.

Giraud, D., K. Klonsky and P. Livingston. 2005. Sample costs for a 500 dairy goat operation: milk for cheese production in the North Coast. University of California Cooperative Extension, Davis. Pp. 15. http://coststudies.ucdavis.edu/files/dairygoatsnc05r.pdf.

Mitcham, S., and A. Mitcham. 2006. Second edition. Meat goats: Their history, management, and diseases. Crane Creek Publications, Sumner, IA. Pp. 288.

Solaiman, S. G. 2006. Outlook for a small farm meat goat industry in California. University of California Small Farm Program, Davis. Pp. 28.

Sheep

American Sheep Industry Association. 2003. SID sheep production handbook. 2002 edition, vol. 7. American Sheep Industry, Inc. Pp. 1060.

Loomis, E. C. 2002. A handbook for raising small numbers of sheep. Third edition. University of California Division of Agriculture and Natural Resources, Oakland. Publication 21389. Pp. 73. http://anrcatalog.ucdavis.edu/Sheep/21389.aspx.

Beekeeping

Avitabile, A., D. Sammataro, and R. Morse. 2006. The beekeeper's handbook. Third edition. Cornell University Press, Ithaca, NY. Pp. 280.

Beekeeping in California. 1987. Out of print, but available online. http://www.beeguild.org/CA_Beekeeping_V2.pdf.

Blackiston, H. 2009. Beekeeping for dummies. Second edition. Wiley, Hoboken, NJ. Pp. 392.

Graham, J., ed. 1992. The hive and the honey bee. Dadant & Sons, Hamilton, IL. Pp. 1324.

Llamas and Alpacas

Kadwell, M., M. Fernandez, H. F. Stanley, R. Baldi, J. G. Wheeler, R. Rosadio, and M. W. Bruford. 2001. Genetic analysis reveals the wild ancestors of the llama and the alpaca. Proc. R. Soc. Long. B 268:2575–2584.

Alpaca Owners and Breeders Association
5000 Linbar Drive, Suite 297
Nashville, TN 37211
Telephone: (800) 213-9522
E-mail: infor@aobamail.com
http://www.alpacainfo.com.

Alpaca Registry, Inc.
4711 Innovation Drive, Suite 160
Lincoln, NE 68521
Telephone: (402) 437-8484
http://www.alpacaregistry.net.

California International Llama Association
1188 Olive Hill Lane
Napa, CA 94558
Telephone: (707) 255-2621
http://www.cal-ila.org.

International Llama Registry
The International Lama Registry
PO Box 8
Kalispell, MT 59903
Telephone: (406) 755-3438
http://www.lamaregistry.com.

Waste Management

American Society of Agricultural and Biological Engineers. 2005. Manure production and characteristics. ASABE Standard D384.2. American Society of Agricultural and Biological Engineers, St. Joseph, MI. Pp. 20.

Small Farm Profiles

Dan Macon
Flying Mule Farm

Jay Ruskey
Good Land Organics

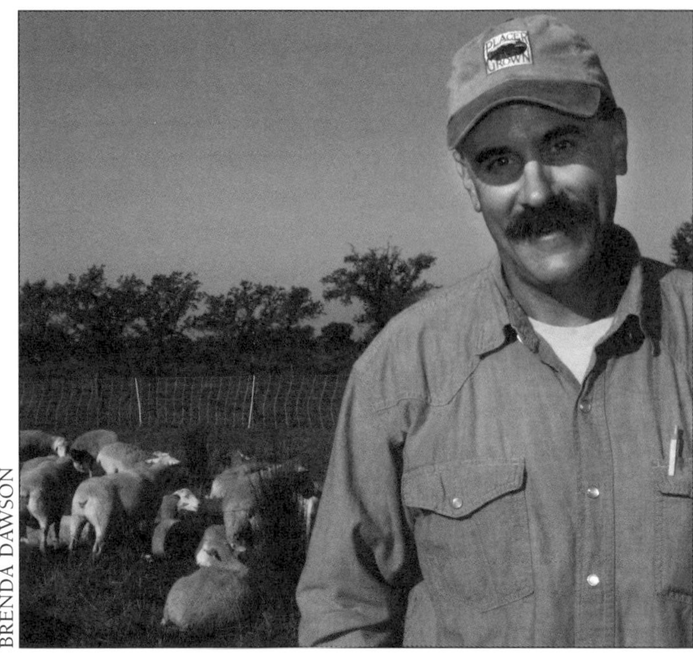

BRENDA DAWSON

SMALL FARM PROFILE:

Dan Macon

Flying Mule Farm

"For me, farming wouldn't be farming if we weren't raising some sort of animal," says Dan Macon, owner and operator—along with his wife and two young daughters—of Flying Mule Farm in the Sierra Nevada foothills.

The family produces pastured chickens and eggs on 3.5 acres at their home and harvests firewood with help from a mule named Frisbee. Their main product is grass-fed lamb, along with grass-fed beef and goat, raised with management-intensive grazing techniques and livestock guardian dogs on approximately 600 acres of leased land.

Lamb is a seasonal product, unlike some other animal enterprises, so Macon is adding stackable enterprises to develop year-round profits, including harvesting firewood, managing land by contract, and raising poultry.

"Like most small farms, we don't enjoy an excess of capital," he says. "What I have to invest, largely, is my time—and if I can find an enterprise that provides some return on my labor on a nearly yearlong basis, then it starts to make economic sense to farm."

Currently the Macons, along with their marketing partners, work with a meat processing plant about 60 miles away where their lambs are slaughtered, cut, and wrapped. Their meat products will soon have custom labels and are sold at farmers markets, to local restaurants, to a small, natural foods retailer, and through the Sierra Foothills Meat Buying Club.

"It's a lot less work to take 150 lambs to the auction, but it's also a lot less rewarding for me, both financially and personally," Macon says of the family's decision to market their products directly to consumers.

He notes that a positive aspect of rapid urbanization can be farmers' increased access to markets and customers. When he sells his own products, Macon can look across the table at his customers and tell them how the lamb was raised.

But he sees the regulatory system as a challenge, especially when it comes to processing meat for small-scale ranchers. He explains that the meat inspection system makes it so consumers do not have to know who raised their meat, since it's the government that vouches for its safety.

"I think the regulatory system is set up to deal with large operations and maintain a sense of anonymity between producers and their consumers," he says. "For small growers, I think that's a real challenge."

Another challenge common to raising animals can be neighbors, who do not always expect to live next door to livestock noises and smells. The family has found that educating neighbors about their farm's animals has for the most part led to acceptance. "While we did have one complaint about nighttime barking from a livestock guardian dog, it never became a right-to-farm issue."

In addition to the lambs and guardian dogs, the family also works with mules who help harvest firewood.

For Macon, using mules as part of his farming operation is a philosophical and economic decision. He cites greater self-sufficiency, lower fossil fuel requirements, quieter working conditions, and a sense of partnership with the animals as some of the mules' main advantages over tractors and other machinery.

"Now we buy hay instead of fertilizer, and process it by mule," he says. "We joke that the 'exhaust' from our 'tractor' is also useful."

—*Brenda Dawson*

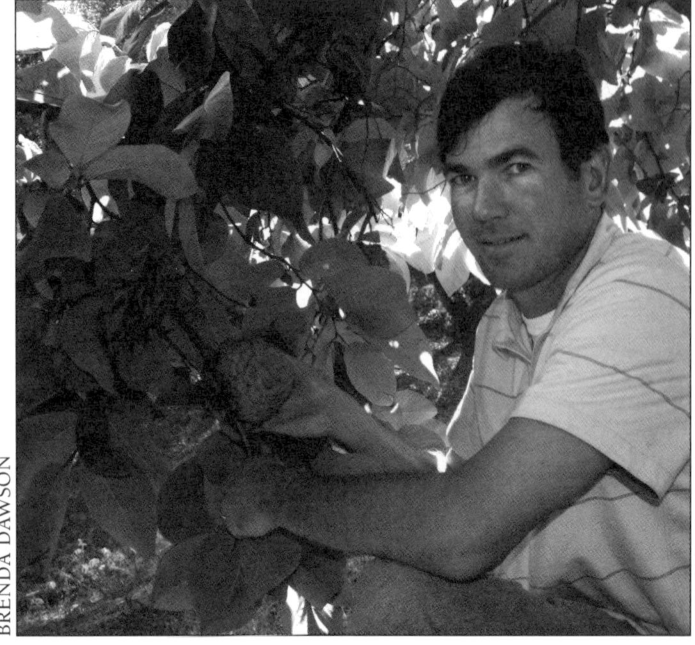
BRENDA DAWSON

SMALL FARM PROFILE:

Jay Ruskey

Good Land Organics

"For years I was very intuitive," says Jay Ruskey, specialty crops farmer with Good Land Organics and Calimoya Brand Cherimoyas, about his choice of crops. "Now to pursue new crops, I look at the risk."

Ruskey has turned his family's 42-acre avocado farm into a diversified, organic mix of subtropical fruits, including cherimoya, passion fruit, strawberry guava, dragon fruit, persimmon, and coffee. He also manages three neighboring farm operations near coastal Goleta, north of Santa Barbara.

"We always have a couple of crops in the hopper, which I think is the diversity that someone needs on this size of farm," he explains. To make decisions about what crops to start growing, Ruskey says he "nerds out" with spreadsheets to weigh the risks of a particular crop over the next 5 years.

"Normally I look at two things—cultivation risk and market risk—on a scale of 1 to 10," he says.

Ruskey cites lychees as an example. He has a trial planting of lychees in partnership with UC Cooperative Extension Farm Advisors. Lychees are a high-value crop with an established ethnic market, so he says their market risk is low, perhaps a score of 2 or 3 on his scale. But producing the fruit consistently is very challenging, so the cultivation risk is high, an 8 or 9.

Before he ventured into new crops, Ruskey's farm was an avocado-only operation, and he still sells avocados as a commodity crop.

"When I took over the farm, they took everything to a major packinghouse in 600-pound bins. But with specialty crops, you're now responsible for how the crop gets distributed and how it is perceived by the customer," he explains. "When I first started, we had to look at the marketing side of things because that's what drives everything."

Ruskey sells many of his subtropical fruits directly to restaurants around the country, to consumers through his online business, and to specialty brokers. For the last 15 years, he's also sold at farmers markets, which he sees as very useful, especially with introducing new crops.

"That's a test market," he says, "so I've been able to look at consumers, how they behave, and their response to my fruit."

To find a new niche crop, Ruskey suggests that the prospective grower become an avid reader, talk with chefs, and call specialty crops brokers. He keeps some new crops a surprise from potential competitors, and suggests that farmers grow crops they actually like. Once you have found a crop and a market for it, then the pressure is on to cultivate the crop.

For rare crops, the production risk can be high. Ruskey says it is important to remember that "planned failure" and mistakes along the way are part of the business plan for rare crops, and he suggests contacting UC Cooperative Extension for assistance.

In addition to deciding what crops to grow, Ruskey suggests that new farmers also decide what size of farm they want to have. During packing season for cherimoyas, his crew is at maximum capacity; even if he doubled the farm's trees and invested in machinery for the packinghouse, the farm might still not be big enough to supply a regional grocery chain.

"It's a dangerous world if you want to jump to a larger size," he says. "I have people who want me to jump to a larger size, but I like the specialty niche. I put a lot of energy into my crops and the people who grow my crops."

—Brenda Dawson

Small Farm Handbook

APPENDIX

Appendix

University of California Cooperative Extension Offices

Office locations and telephone numbers may change. For up-to-date contact information, check the County Office Web sites listed below or visit UC Division of Agriculture and Natural Resources online at http://ucanr.org.

Alameda County
1131 Harbor Bay Parkway, Suite 131
Alameda, CA 94502
Telephone: (510) 567-6812
Fax: (510) 748-9644
http://cealameda.ucdavis.edu

Amador County
12200-B Airport Road
Jackson, CA 95642-9527
Telephone: (209) 223-6482
Fax: (209) 223-3279
http://ceamador.ucdavis.edu

Butte County
2279-B Del Oro Avenue
Oroville, CA 95965
Telephone: (530) 538-7201 (call ahead for ADA assistance)
Fax: (530) 538-7140
http://cebutte.ucdavis.edu

Calaveras County
891 Mountain Ranch Road
San Andreas, CA 95249
Telephone: (209) 754-6477
Fax: (209) 754-6472
http://cecalaveras.ucdavis.edu

Colusa County
PO Box 180, 100 Sunrise Boulevard, Suite E
Colusa, CA 95932
Telephone: (530) 458-0570
Fax: (530) 458-4625
http://cecolusa.ucdavis.edu

Contra Costa County
75 Santa Barbara Road, 2nd Floor
Pleasant Hill, CA 94523-4215
Telephone: (925) 646-6540
Fax: (925) 646-6708
http://cecontracosta.ucdavis.edu

El Dorado County
311 Fair Lane
Placerville, CA 95667
Telephone: (530) 621-5502
Fax: (530) 642-0803
http://ceeldorado.ucdavis.edu

Fresno County
1720 South Maple Avenue
Fresno, CA 93702
Telephone: (559) 456-7285
Fax: (559) 456-7575
http://cefresno.ucdavis.edu

Glenn County
PO Box 697, 821 E. South Street
Orland, CA 95963
Telephone: (530) 865-1107
Fax: (530) 865-1109
http://ceglenn.ucdavis.edu

Humboldt–Del Norte Counties
5630 South Broadway
Eureka, CA 95503-6998
Telephone: (707) 445-7351
Fax: (707) 444-9334
http://cehumbolt.ucdavis.edu

Imperial County
1050 East Holton Road
Holtville, CA 92250-9615
Telephone: (760) 352-9474
Fax: (760) 352-0846
http://ceimperial.ucdavis.edu

Inyo-Mono Counties
207 W. South Street
Bishop, CA 93514
Telephone: (760) 873-7854
Fax: (760) 873-7314
http://ceinyo-mono.ucdavis.edu

Kern County
1031 South Mount Vernon Avenue
Bakersfield, CA 93307
Telephone: (661) 868-6200
Fax: (661) 868-6208
http://cekern.ucdavis.edu

Kings County
680 North Campus Drive – A
Hanford, CA 93230
Telephone: (559) 582-3211
Fax: (559) 582-5166
http://cekings.ucdavis.edu

Lake County
883 Lakeport Boulevard
Lakeport, CA 95453
Telephone: (707) 263-6838
Fax: (707) 263-3963
http://celake.ucdavis.edu

Lassen County
707 Nevada Street
Susanville, CA 96130
Telephone: (530) 251-2601
Fax: (530) 251-2666
http://celassen.ucdavis.edu

Los Angeles County
4800 E. Cesar E. Chavez Avenue
Los Angeles, CA 90022
Telephone: (323) 260-2267
Fax: (323) 260-5208
http://celosangeles.ucdavis.edu

Madera County
328 Madera Avenue
Madera, CA 93637
Telephone: (559) 675-7879
Fax: (559) 675-0639
http://cemadera.ucdavis.edu

Marin County
1682 Novato Boulevard, 150B
Novato, CA 94947
Telephone: (415) 499-4204
Fax: (415) 499-4209
http://cemarin.ucdavis.edu

Mariposa County
5009 Fairgrounds Road
Mariposa, CA 95338-9435
Telephone: (209) 966-2417
Fax: (209) 966-5321
http://cemariposa.ucdavis.edu

Mendocino County
890 N. Bush Street
Ukiah, CA 95482
Telephone: (707) 463-4495
Fax: (707) 463-4477
http://cemendocino.ucdavis.edu

Merced County
2145 Wardrobe Avenue
Merced, CA 95340-6445
Telephone: (209) 385-7403
Fax: (209) 722-8856
http://cemerced.ucdavis.edu

Modoc County
202 West Fourth Street
Alturas, CA 96101
Telephone: (530) 233-6400
Fax: (530) 233-3840
http://cemodoc.ucdavis.edu

Monterey County
1432 Abbott Street
Salinas, CA 93901
Telephone: (831) 759-7350
Fax: (831) 758-3018
http://cemonterey.ucdavis.edu

Napa County
1710 Soscol Avenue, Suite 4
Napa, CA 94559-1315
Telephone: (707) 253-4221
Fax: (707) 253-4434
http://cenapa.ucdavis.edu

Orange County
1045 Arlington Drive
Costa Mesa, CA 92626
Telephone: (714) 708-1606
Fax: (714) 708-2754
http://ceorange.ucdavis.edu

Placer-Nevada Counties
11477 E Avenue
Auburn, CA 95603
Telephone: (530) 889-7385
Fax: (530) 889-7397
http://ceplacer.ucdavis.edu

Plumas-Sierra Counties
208 Fairgrounds Road
Quincy, CA 95971
Telephone: (530) 283-6270
Fax: (530) 283-6088
http://ucce-plumas-sierra.ucdavis.edu

Riverside County
21150 Box Springs Road, 202
Moreno Valley, CA 92557
Telephone: (951) 683-6491
Fax: (951) 788-2615
http://ceriverside.ucdavis.edu

Sacramento County
4145 Branch Center Road
Sacramento, CA 95827-3898
Telephone: (916) 875-6913
Fax: (916) 875-6233
http://cesacramento.ucdavis.edu

San Benito County
649 San Benito Street, 115
Hollister, CA 95023
Telephone: (831) 637-5346
Fax: (831) 637-7111
http://cesanbenito.ucdavis.edu

San Bernardino County
777 East Rialto Avenue
San Bernardino, CA 92415
Telephone: (909) 387-2171
Fax: (909) 387-3306
http://cesanbernardino.ucdavis.edu

San Diego County
5555 Overland Avenue, Building 4
San Diego, CA 92123-1219
Telephone: (858) 694-2845
Fax: (858) 694-2849
http://cesandiego.ucdavis.edu

San Joaquin County
420 South Wilson Way
Stockton, CA 95205
Telephone: (209) 468-2085
Fax: (209) 462-5181
http://cesanjoaquin.ucdavis.edu

San Luis Obispo County
2156 Sierra Way, Suite C
San Luis Obispo, CA 93401
Telephone: (805) 781-5940
Fax: (805) 781-4316
http://cesanluisobispo.ucdavis.edu

San Mateo–San Francisco Counties
80 Stone Pine Road #100
Half Moon Bay, CA 94019
Telephone: (650) 726-9059
Fax: (650) 726-9267
http://cesanmateo.ucdavis.edu

Santa Barbara County
2156 Sierra Way, Suite C
San Luis Obispo, CA 93401
Telephone: (805) 781-5940
Fax: (805) 781-4316
http://cesantabarbara.ucdavis.edu

Santa Clara County
1553 Berger Drive, Building 1
San Jose, CA 95112
Telephone: (408) 282-3110
Fax: (408) 298-5160
http://cesantaclara.ucdavis.edu

Santa Cruz County
1432 Freedom Boulevard
Watsonville, CA 95076-2796
Telephone: (831) 763-8040
Fax: (831) 763-8006
http://cesantacruz.ucdavis.edu

Shasta County
1851 Hartnell Avenue
Redding, CA 96002-2217
Telephone: (530) 224-4900
Fax: (530) 224-4904
http://ceshasta.ucdavis.edu

Siskiyou County
1655 South Main Street
Yreka, CA 96097
Telephone: (530) 842-2711
Fax: (530) 842-6931
http://cesiskiyou.ucdavis.edu

Solano County
501 Texas Street
Fairfield, CA 94533-4498
Telephone: (707) 784-1317
Fax: (707) 429-5532
http://cesolano.ucdavis.edu

Sonoma County
133 Aviation Boulevard, Suite 109
Santa Rosa, CA 95403
Telephone: (707) 565-2621
Fax: (707) 565-2623
http://cesonoma.ucdavis.edu

Stanislaus County
3800 Cornucopia Way
Modesto, CA 95358
Telephone: (209) 525-6800
Fax: (209) 525-6840
http://cestanislaus.ucdavis.edu

Sutter-Yuba Counties
142 Garden Highway, Suite A
Yuba City, CA 95991-5512
Telephone: (530) 822-7515
Fax: (530) 673-5368
http://cesutter.ucdavis.edu

Tehama County
1754 Walnut Street
Red Bluff, CA 96080
Telephone: (530) 527-3101
Fax: (530) 527-0917
http://cetehama.ucdavis.edu

Trinity County
PO Box 490, 3000 Highway 3
Trinity County Fairgrounds
Hayfork, CA 96041
Telephone: (530) 628-5495
Fax: (530) 628-1945
http://cetrinity.ucdavis.edu

Tulare County
4437 S. Laspina Street, Suite B
Tulare, CA 93274
Telephone: (559) 685-3303
Fax: (559) 685-3319
http://cetulare.ucdavis.edu

Tuolumne County
2 S. Green Street
Sonora, CA 95370
Telephone: (209) 533-5695
Fax: (209) 532-8978
http://cetuolumne.ucdavis.edu

Ventura County
669 County Square Drive, 100
Ventura, CA 93003-5401
Telephone: (805) 645-1451
Fax: (805) 645-1474
http://ceventura.ucdavis.edu

Yolo County
70 Cottonwood Street
Woodland, CA 95695
Telephone: (530) 666-8143
Fax: (530) 666-8736
http://ceyolo.ucdavis.edu

INDEX

T